A Manual of Practical Zoology

Part - 1

Biodiversity, Cell Biology, Genetics and Developmental Biology

AF529877

The Authors

Dr. Madan Mohan Trigunayat obtained his Master's degree standing first among first divisioners at Department of Zoology, University of Rajasthan, Jaipur in 1983. He joined teaching assignment in Govt. of Rajasthan in 1984. He has worked as Research Entomologist in USDA (United States Department of Agriculture) scheme. He has got published more than 80 research papers in leading journals and also attended International conferences (held in India) and many National conferences and seminars. He has been included in Entomologist's Profile of India in 1997 and Reference Asia in 1995 and Learned Asia in 1996. He was awarded Summer Research Fellowship jointly from three National Academies of Sciences, India in 2011. He organized a National conference on Forest and Wildlife : Present status and future needs in 2013. He is teaching post graduate and graduate classes for more than 34 years. He is the author of 13 other books on zoology. His present research interests are Insect Ecology and Biodiversity and Agricultural Ornithology.

Kritika Trigunayat, a Gold Medalist in Graduation from University Maharani's College, Jaipur in 2011 and also a Gold Medalist at Post Graduation in Zoology with specialization in Entomology, from the same University in 2013. She was also selected for Student Summer Fellowship scheme of three National Academies of Sciences, India in 2012 and has worked on the Eco biology of an insectivorous bird, Ashy Wren Warbler, Prinia Socialis at APJ Satya University, Gurgaon. She was awarded Inspire Fellowship from the Department of Science and Technology, New Delhi in 2015.

Presently she is pursuing research leading to Ph.D at Department of Zoology, University of Rajasthan, Jaipur under the supervision of Prof. Jaimala Sharma on Coleopteran Diversity in Keoladeo National Park, Bharatpur.

A Manual of Practical Zoology

Part - 1

Biodiversity, Cell Biology, Genetics and Developmental Biology

Authors

M.M. Trigunayat, Ph.D., F.E.S.I., F.Z.S.I.

Head, Deptt. of Zoology
Govt R.D. Girls (PG) College, Bharatpur

&

Kritika Trigunayat

(M.Sc. Gold Medalist)
DST-Inspire Fellow
Department of Zoology,
University of Rajasthan, Jaipur

SCIENTIFIC
PUBLISHERS

Published by

SCIENTIFIC PUBLISHERS (INDIA)
5 A, New Pali Road
P.O. Box 91
Jodhpur - 342 001 INDIA
E-mail: info@scientificpub.com
Website: http://www.scientificpub.com

Print: 2019

All rights reserved. No part of this publication or the information contained herein may be reproduced, adapted, abridged, translated, stored in a retrieval system, computer system, photographic or other systems or transmitted in any form or by any means, electronic, mechanical, optical, digital, by photocopying, recording or otherwise, without written prior permission from the publisher. Any breach will attract legal action and prosecution without further notice.

Disclaimer: While every effort has been made to avoid errors and omissions, this publication is being sold and marketed on the understanding and presumption that neither the editors (or authors) nor the publishers nor the printers would be liable in any manner whatsoever, to any person either for an error or for an omission in this publication, or for any action to be taken on the basis of this work. Any inadvertent discrepancy noted may be brought to the attention of the publisher, for rectifying it in future editions, if published.

This book contains information obtained from authentic and highly regarded sources. Reasonable efforts have been made to publish reliable data and information, but the editors and publisher cannot assume responsibility for the validity of all materials or the consequences of their use. The editors and publisher have attempted to trace and acknowledge the copyright holders of all material reproduced in this publication and apologize to copyright holders if permission and acknowledgement to publish in this form have not been obtained. If any copyright material has not been acknowledged please write and let us know so that we may rectify it.

Trademark Notice: Publications or corporate names may be trademarks, and are used only for identification and explanation in bonafide intent without intent to infringe.

ISBN: 978-93-88449-07-6 (Paperback)
978-93-88449-08-3 (E-book)

© 2019, Trigunayat, M.M.

Visit the Scientific Publishers (India) website at
http://www.scientificpub.com
Printed in India

PREFACE

The present book entitled `**A Manual of Practical Zoology**' (Part 1) is written according to new and revised syllabi of UGC adopted for B.Sc. Pt. 1 of various universities including Rajasthan University, Jaipur and its affiliated colleges and some other universities started new in Rajasthan and elsewhere. It consists of 7 Units. Unit first envisage working of microscopes, culture methods and museum specimens of invertebrates. The specimens have been described in a unique style as the information on vital aspects of Eco biology of the type concerned *viz.* Habit and habitat, Distribution, Characters of identification, Special features, Biological importance and Economic importance (if any) have been given. These are the points on which the marks of the spots are distributed. Standard culture methods for protozoans have been described. Note on vermiculture is also given as it is gaining popularity now a days. Unit second covers the anatomy. Easy drawings have been incorporated and *in situ* figures of various anatomical structures have been given. Unit third covers permanent slides and fourth with exercises in cell biology. Fifth and Sixth Units are on genetics and developmental biology respectively. Unit seventh covers simple exercises on Biodiversity. As per the guidelines from UGC, many universities have discontinued dissections and anatomy has been replaced by Biodiversity based exercises. M.S. Braj University has added Biodiversity study in Practical curriculum. The language is easily understandable and the figures are easy to draw. The overall presentation of the text is unique. We hope it shall be preferred by the students and teachers. Additional informations wherever needed have been given.

Teachers and friends have been helpful in one way or other. We express our deep sense of gratitude to them. We thank Shri Pawan Kumar and Tanay Sharma of Scientific Publishers (India), Jodhpur for bringing this Manual in very short time.

Suggestions of teachers, colleagues and students are invited.

Authors

Bharatpur
10.10.2018

CONTENTS

Unit - 1

Microscopic Technique and Museum Specimens

1. ORGANIZATION AND WORKING OF MICROSCOPES

Compound Microscope

Compound microscope is used to magnify very small objects. It has two units. The *mechanical unit* and *optical unit.*

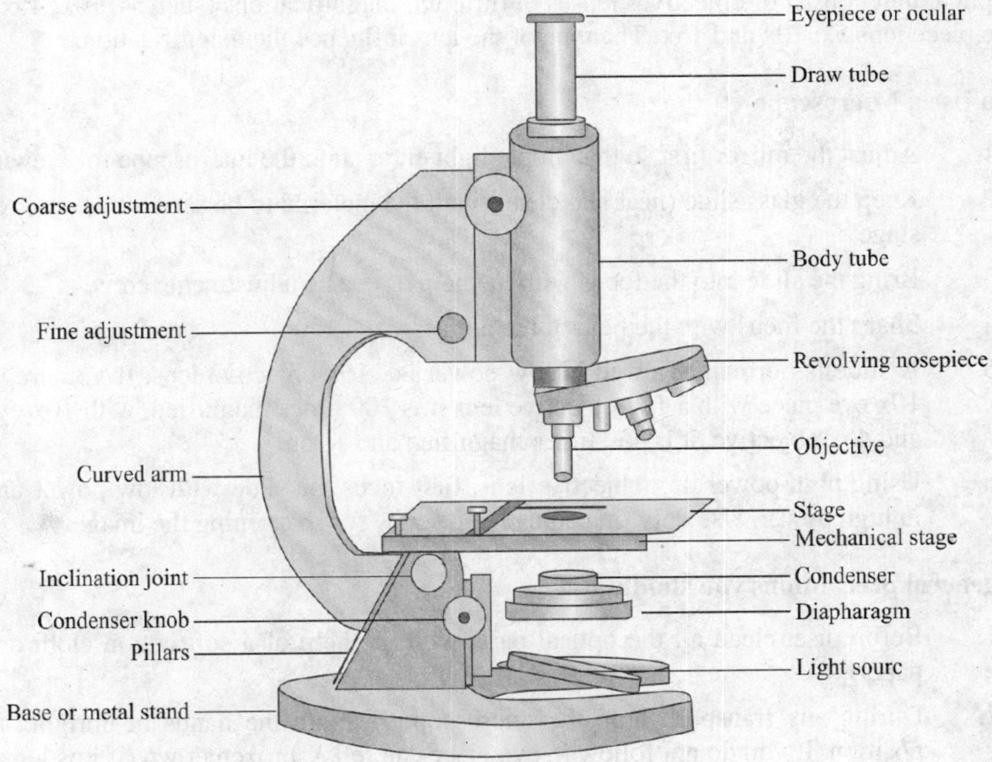

Fig. 1.1a: The compound microscope

Mechanical unit consist of the different part of the microscope *viz.* Base, Arm Joint, Stage, Clips, Diaphragm. Body, Nose piece, Coarse and Fine adjustment screws (Figure 1.1).

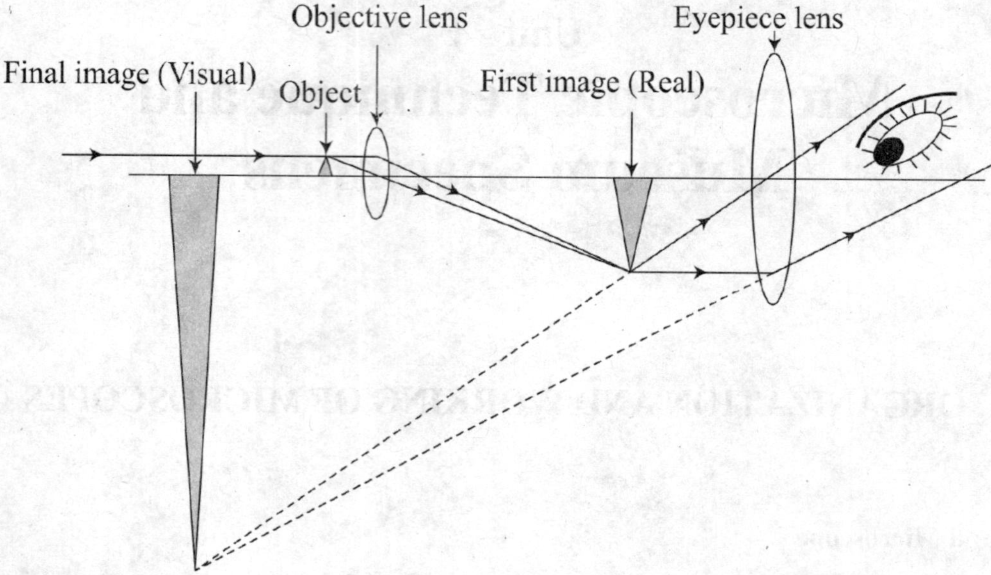

Fig. 1.1b. Working of Compound Microscope

Optical unit consist of objectives lenses of different magnifications (such as, 10x, 45x, 100x) and eye piece lens 5x, 10x and 15x. The type of the lens influence the magnification.

How to Use a Microscope

1. Adjust the mirror first so that ample light enters into the microscope for viewing.
2. Keep the glass slide (neat and clean) with the material to be seen, at the centre of the stage.
3. Bring the slide into the focus with the help of coarse adjustment screw.
4. Sharp the focus with the help of the adjustment screw.
5. A slide is normally viewed in low power i.e. 10x objective lens. If you are using a 10x eye piece with a 10x objective lens it is 100 times magnified, with 10x eyepiece and 45x objective, it is 450 times magnified and so on.
6. Using high power 45x objective lens, first focus the slide with low power and then in high power. Use only fine adjustment screw for sharpening the image.

What general precautions you should use

1. Before use, clean all the optical parts with the help of a soft cotton cloth or tissue paper.
2. During any transport, hold the microscope by both the hands in upright, straight position. If you do not follow it, eye piece can fell & broken down (if it is loose)

3. Always focus the microscope in low power first and then only in high power. After use change it in low power.
4. Always use fine adjustment with high power. Do not use coarse adjustment screw frequently with high power.
5. Keep the microscope covered in its case when not being used.

Dissecting microscope

It is used when high magnification is not required, 5 x, 10x and 15x lenses are generally used. A dissecting microscope consist of the following parts.

1. Base 2. Stand 3. Vertical limb 4. Folded arm 5. Stage and Mirror.

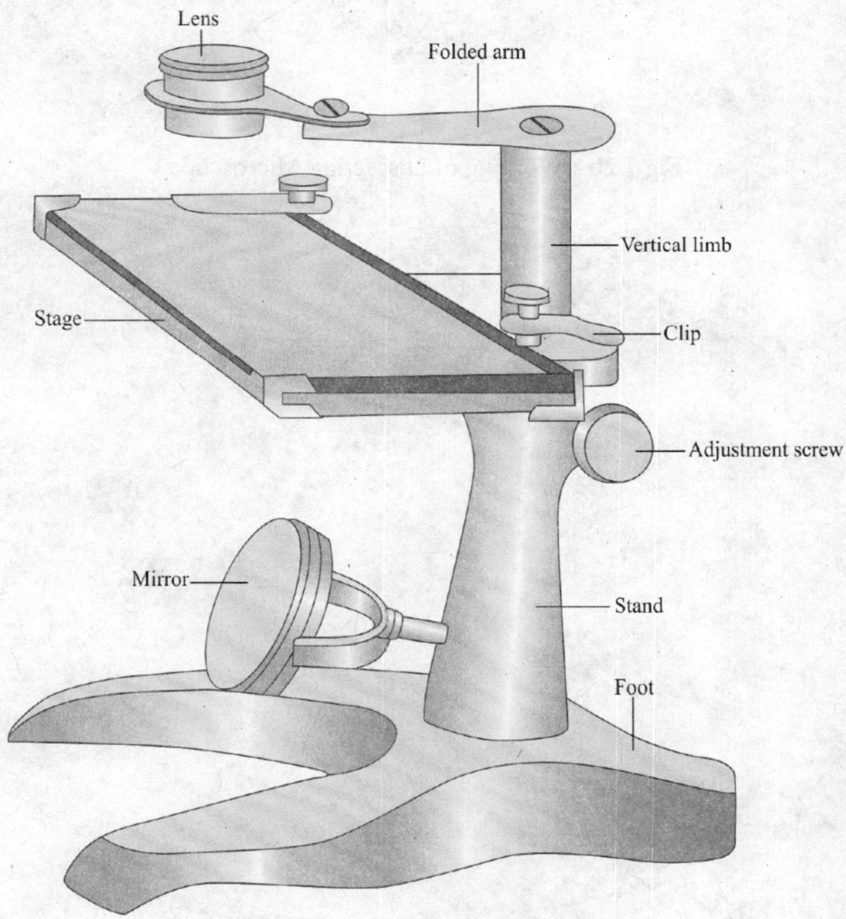

Fig. 1.2a : Dissecting microscope

Working of dissecting microscope

1. Place the slide (to be seen) on the glass stage and bring the lens over to it with the help of folded arm. Adjust the light with the help of a mirror attached lower end of the stand.

2. Focus the slide with the help of vertical limb (moved up ad down). Vertical limb is moved, by adjustment screws.

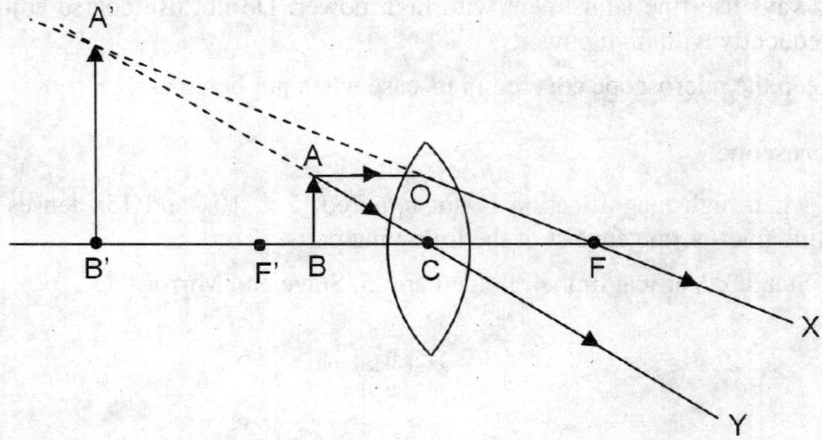

Fig 1.2b : Working of Dissecting Microscope

2. GENERAL METHODS OF MICROSCOPICAL PERMANENT PREPARATIONS

NARCOTIZATION

It is done either with ether vapours or chloroform. Ethyl acetate is generally used for insects. Carbon tetra chloride is also used. You can also use dry ice/liquid CO_2. The purpose is to inactivate the animal immediately.

WASHING

It is done with the tap water (of Formalin preserved material)

DEHYDRATION

It is done by passing the tissue in a series of alcohol grades 30%, 50%, 70%, 90%, absolute alcohol, clearing is done in xylene and the tissue is mounted in DPX or Canada Balsam.

Staining is done after 70% alcohol if the eosin is prepared in it. Excess stain is removed by destaining by passing back the tissue in 70% alcohol or acid alcohol. The tissue is mounted in DPX because the refractive index of the DPX is nearly equal to the refractive index of glass (=1) (of which the slide is made of)

Dehydration is always done in series i.e. 30%, 50%, 70%, 90% and absolute alcohol because the water is gradually removed. If this series is not followed then the hypertonic solution will distort the shape of the tissue we are studying therefore we cannot put the tissue directly in 90% alcohol or so.

Fixatives

The chemicals which fix the tissues or cells or animal so that they can be studied for longer duration in the laboratories are generally called fixative. The tissue gets harden and the contents are preserved. Larger tissues are keep in fixative for longer time than smaller tissues which requires less time. A number of chemicals are used as fixative e.g. *Bouin's fluid, Formalin, Carnoy's fluid, Corrosive sublimate, FAA, Carl's fluid etc.*

Table 1. Some general fixatives

S.No.	Name of the fixative	Composition	Use and Time allowed for fixation
1	Bouin's Fluid	Glacial Acetic Acid, HCHO, Picric Acid	It is a general fixative and allow the tissue for 24 hrs.
2	Formalin	HCHO 40% and water	Most commonly used fixative for vertebrates. It is diluted as 10% or 5%. Animals are kept for long periods.

S.No.	Name of the fixative	Composition	Use and Time allowed for fixation
3	Zenker's solution	$K_2Cr_2O_7$ $HgCl_2$, Na_2SO_4. Acetic acid (glacial) and water	Good for histological work but it is not very stable.
4	FAA	90% Ethyl alcohol, HCHO 40%, Acetic acid (glacial) and distilled water	It is a good preservative for animals. The tissues can be stored for years together. Nematodes are fixed in it.
5	Carnoy's fluid	Acetic acid (glacial) 95% alcohol, CCl_4	Good for histological studies. The tissues are fixed at least for a day.
6	Carl's fluid	90% Ethyl alcohol, HCHO 40% Distilled water and Acetic acid (glacial)	Good fixative for insects. Glycerine drops are added for hard bodied insects.
7	Kleinberg's fixative	H_2SO_4, picric Acid	It is used for chick embryos and small marine organisms. The tissue is kept for atleast 10-12 hrs.

Preparation of Bouin's fixative

Bouin's (Aqueous)

Take 5 ml of Glacial acetic acid, 25 ml of 40% Formaldehyde (HCHO) and 75 ml of saturated (in distilled water) Picric acid. Mix them and shake and store.

Bouin's (Alcoholic)

As above except the picric acid is saturated in alcohol.

Preparation of Formalin

Dilute 40% HCHO (Formaldehyde) to 10% or 5%. Formalin is 40% solution of formaldehyde gas dissolved in water.

Stains

To study a tissue or animal in the laboratory we require to differentiate the various structures. It is done by chemical which provide colour to the tissue. They are categorized into *vital* and *non-vital stains.* Non vital stains cannot be used for living cells whereas the vital stains are used to colour the living cells. *Methylene blue, Congo red, Neutral red* and *Janus green* are commonly used vital stains. Commonly used non vital or simple stains includes Acetocarmine, Aceto-orcein, Haematoxylin, Eosin and Borax carmine.

Table 2 : Some Vital and Non vital stains

S.No.	Name of the Stain	Composition	Use
	Vital		
1	Methylene blue	Methylene blue powder and distilled water	Small organisms cell and cytoplasmic particles, nucleus are stained.
2	Congo red	Congo red powder and water (distilled).	
3	Neutral red	Neutral red powder and alcohol	To stain nucleus.
4	Janus green	Janus green powder + distilled water	To stain Mitochondria.
	Non-Vital		
1	Aceto-carmine	Carmine powder + Acetic acid (glacial) and Distilled water	Nucleus, Protozoa, Meiosis, Mitosis, Polytene chromosomes.
2	Aceto orcein	Orcein powder, Acetic acid (glacial) and Distilled water	Polytene chromosomes, Nucleus.
3	Borax carmine	Borax, carmine powder, Alcohol and Distilled water	Nucleus (pink)
4	Crystal violet	Dye crystal violet, Ethyl alcohol	Bacteria
5	Basic fuchsin	Dye fuchsin, Ethyl alcohol	Bacteria
6	Haematoxylin	Haematoxylin, Absolute alcohol, Glycerol, Glacial Acetic acid, Aluminium potassium sulphate and Distilled water	Nuclear stain
7	Iodine	KI, Iodine & distilled water	To stain flagella cilia & nucleus.
8	Picro indigo carmine	Picric acid 90% Alcohol and Indigo	To stain chitinous material.

Preparation of Eosin

Take 1 gm eosin powder and dissolve it into 100 ml distilled water (*Aquous eosin*).

Take 1 gm eosin powder and dissolve it into 100 ml 70% alcohol (*Alcoholic eosin).*

Preparation of Haematoxylin

Ehlrich's haematoxylin

Take 2 gm Haematoxylin and dissolve it into 100 ml ethyl alcohol. It is *Solution A.*

Also take 100 ml Glycerol and distilled water each, 10 ml. of Glacial acetic acid and 3 gm of aluminium potassium sulphate. It is *Solution B.* Mix both A and B solutions and shake well thoroughly. Allow this solution to mature for two weeks.

Delafield Haematoxylin

Take 4 gm Haematoxylin and dissolve it into 25 ml absolute alcohol. It is *Solution A*.

Add to it 400 ml. of saturated aqueous solution of ammonium alum. This solution is allowed to expose to light at least for two days. Now add methyl alcohol and glycerine (100 ml. each) and mix thoroughly. Allow it to ripen for at least 60 days or so.

Preparation of Aceto carmine and Aceto orcein

Take 2 gm *carmine* powder and dissolve it into 100 ml. of 45% boiling acetic acid. Again heat for 10 minutes. Cool and filter it. Label it as Acetocarmine.

Aceto orcein

Take 2 gm *orcein powder* and dissolve it into 100 ml. of 45% boiling acetic acid. Again heat for 10 minutes. Cool and filter it. Label it as Aceto orcein.

Preparation of Borax carmine

Take 2.5 gm, carmine powder and 4.0 gm of Borax. Mix them and dissolve into 1000 ml of water. Add to it 1000 ml alcohol. Allow to stand for 4 days and filter it.

Common Reagents

Various solutions and reagents are prerequisite for many experiments performed in the laboratories. Some common reagents includes, normal saline, Ringer's solution. Mayer's albumin, acid water and acid alcohol.

Preparation of Normal Saline

0.7% Nacl

Take 0.7 gm of NaCl and dissolve it into 100 ml of distilled water. It is used for vertebrates.

0.9% Nacl

Take 0.9 gm NaCl and dissolve it into 100 ml of distilled water. It is used for mammals.

Ringer's Solution

Dissolve the following in distilled water.

Sodium Chloride	-	0.65 gm
Sodium Carbonate	-	0.62 gm
Calcium Chloride	-	0.03 gm
Potassium Chloride	-	0.025 gm
Distilled Water	-	100 ml

It is used for insects.

Mayer's Albumin

In the preparation of permanent slides of whole mounts of Protozoans, it is used on the slide so that the protozoa may not wash off. In section cutting also (microtomy) it is used to adhere the section of the tissue on the slide so that during dehydration and staining process the tissue is not washed away.

Take 50 ml egg albumin and glycerine each. Add a crystal of thymol. Mix them, Dilute 3 drops of this solution to 6 ml of distilled water and use whenever required.

3. COLLECTION AND CULTURE METHODS

Undergraduate zoology students are advised to visit any pond or lake or river in their area. The sample of water and under water vegetation can be collected from such natural bodies and examined in the laboratory for the presence of Amoeba, Paramecia, Euglena, Daphnia and Cyclops. They are abundantly present in such waters. The samples preferably be collected early in the morning or late in the evening because the planktons are abundantly found on the surface. Under water vegetation including Chara, decaying leaves, Lily, Sphagnum Sedges, Ceratophyllium etc. can be collected easily. Green scums on the surface of water should also be brought to the laboratory and examined under the guidance of teacher.

Collection and Culture of Amoeba

Place the collected samples given above, in a shallow glass dish or earthenware and keep covered to prevent evaporation of water. Treated water should not be used as the most protozoa are very sensitive to changes in water quality. If tap water is used it should be kept for some days in open vessel.

Take distilled water in a glass trough or finger bowls. Add to it boiled hay (few straws). Inoculate with Euglena, diatoms and algae. This culture thrives well, afterwards inoculate with plenty of amoeba isolated from the natural pond samples. When the culture is established add few straws of boiled or dry hay at intervals of few days. If water is evaporated, add distilled water. If culture do not exhibit good number of amoeba then reinoculate it. Keep the culture in shade or diffused light.

Pure Culture

Take finger bowls and filled it more than 1/2 with distilled water. Add 5-6 crushed grains of rice or wheat. Inoculate with amoeba isolated from the culture (take material from the bottom of the bowl with the help of a pipette). At a regular interval of 7 days, add distilled water (to compensate evaporated water) and few kernels of rice.

Culture of Euglena

The water enriched in organic matter is best suited for Euglena. Boil the manure (dung) collected from a cow or horse in distilled water and allow to stand for at least two days and then inoculate with Euglena isolated from the pond water. (Euglena are plenty in green scum on pond surface). Besides this, Euglena can be cultured by using boiled grains of rice. Add 6-7 grains (boiled) in a finger bowl 2/3 filled with distilled water. Allow it till it develop bacterial scum. Inoculate with Euglena (pure culture). You will be able to found good number of Euglena after $1\frac{1}{2}$ - 2 months because Euglena cultures flourish slowly. In old cultures they are encysted on the surface.

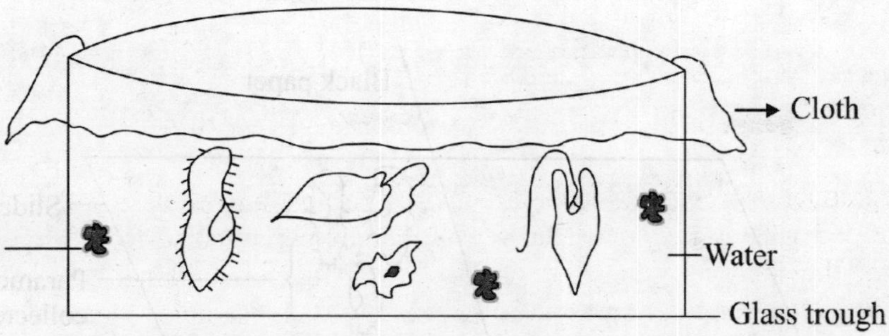

Fig. 1.3 : Culture of Euglena, Paramecium & Amoeba.

Culture of Paramecium

Take a finger bowl or glass trough 2/3 filled with distilled water. Add boiled hay straws. Allow few days to develop bacterial scum. Inoculate with the paramecium isolated from the samples.

Wheat bran (daliya or crushed wheat grains can also be used.)

Behavioural Response of Paramecium

Trial and Error Learning

Paramecium learns through trial and error. It detects the water quality continuously. If some chemical is mixed in the water; it rest on its posterior end and moves the anterior end in a circular fashion so as to avoid the source and change the direction. This is repeated if it encounters the same stimuli.

Avoidance Behaviour

Paramecium exhibit avoidance behaviour if some solid object comes to its way. It change the direction of movement.

Response to Touch

Procedure - Take a drop from the culture and place on the clean glass slide. Focus in the microscope and touch the anterior end with the help of fine needle and observe the reaction.

Result - Paramecium shows strong negative reaction by moving in opposite direction.

Inference - Negative thigmotaxis.

Response to Light

Procedure - Draw some water in a petridish from the culture jar with the help of a dropper. Cover the half petridish with carbon paper or any other black paper or cloth. Now use a table lamp & illuminate it (25 w) and observe the paramecia under stereoscopic binocular.

Result - Paramecia are collected under the black paper.

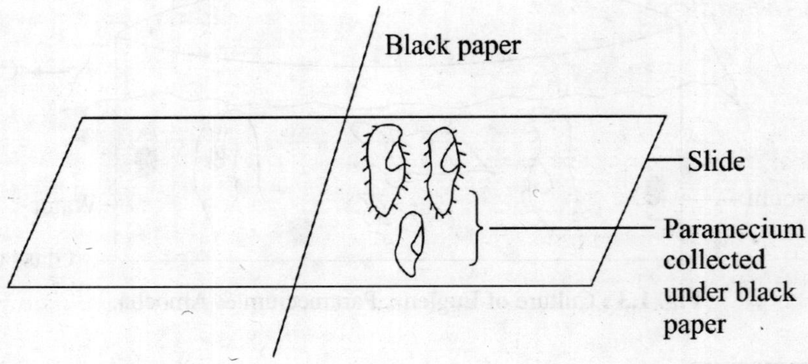

Fig. 1.4 : Response to light

Inference - Negative phototactic movement (They also show positive response in diffused light).

Response to Acid

Procedure- Take few drops of the culture on a clean slide and add a drop of acetic acid. Observe the reaction in microscope.

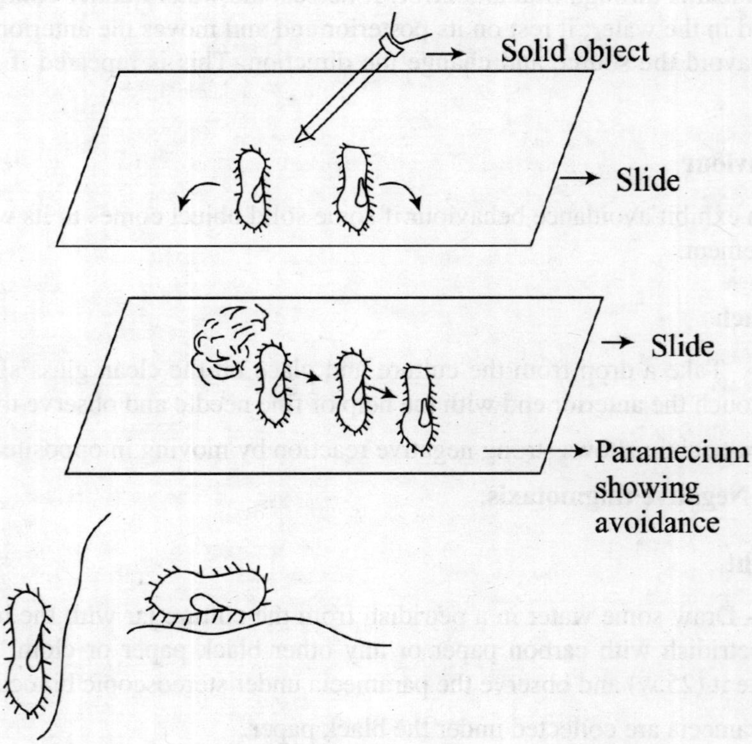

Fig. 1.5 : Response to chemicals, solid object and thread.

Result - The Paramecium avoids and change the direction.

Inference - *Negative chemotaxis.*

Collection and Culture of Planaria

Hyman (1925) suggested baiting for the collection of Planaria. Small lump or piece of meat is placed in shallow waters at the edge of the ponds, streams etc. where they occur. After sometimes they gather over the piece of meat or liver and can be collected in the jar.

They can be collected from the surface of the stones submerged in swiftly flowing water of dams, streams, lakes, ponds etc. Stones are removed from the water ad lower surface is scraped off into dishes of water. Submerged vegetation can also be taken to the laboratory for the search of Planaria. Allow the vegetation to start decaying and the Planaria will collect on the surface of the water and the sides of the container, they can be shifted to finger bowls 3/4 filled with water.

Culture

Temperature 24ºC

Individual finger bowls with Planaria are kept in darkened place. Small piece of liver is given as food. They can also be feed on tubificid worms. To rear new Planaria, cut the posterior part of the body and allow for regeneration. Almost in a month they attained the adult shape.

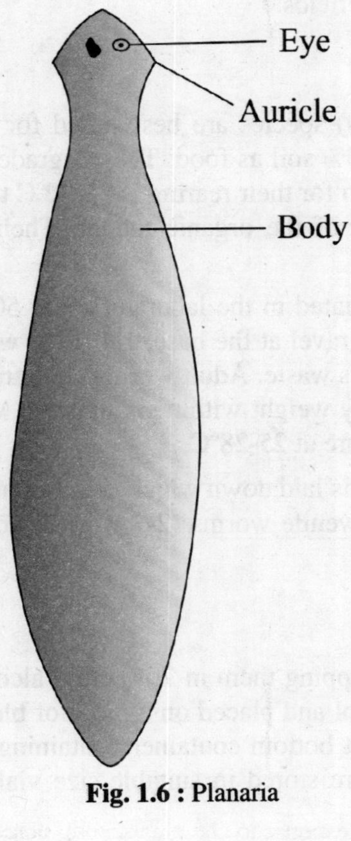

Fig. 1.6 : Planaria

Collection and Culture of Earthworm

The mass culture of earthworms for commercial use is called ***vermiculture***. Vermicomposting is gaining popularity now-a-days. ***Vermicomposting*** deals with biodegradation of waste biomass with the help of Earthworms.

The earthworms are collected from the garden soil during night. They come to the surface of the soil after rains. During rainy season the soil is properly soaked in the water. The worms come out of the soil during night. By using flash light, they can be collected manually. They can be left over night in a pot containing loamy soil and debris of dead leaves.

The best method for collecting earthworms is by digging soil with showel or spade. They can be collected from the following ecological niches viz.

(a) litter
(b) kitchen drainage
(c) manure heaps
(d) different soils
(e) margins of fresh water bodies
(f) pastures
(g) grassland
(h) forest and cultivated fields.

Culture

Epigeic (surface earthworms) species are best suited for vermi composting as they intake 90% of the organic matter and 10% soil as food. They degrade the organic material speedly into *vermicompost.* The ideal condition for their rearing is 25-30ºC temperature and 40-50% humidity. They reproduce within the heap of the organic matter. Their faecal matter is like boiled tea granules.

The culture can even be initiated in the laboratory in a 50 x 25 x 15 cm, wooden box, 1/4 filled with soil, a small layer of gravel at the base, 1/4 filled with saw dust, rice bran, straw; 1/4 cow dung or any other nitrogenous waste. Adult worms (100 are allowed to grow. They can grow to 115-120 times their initial body weight within six months. Moisture conditions of the soil are maintained 15-20% and temperature at 25-28ºC.

After copulation, the cocoon is laid down which contains around 20 eggs. It takes three week to develop the cocoon and the juvenile worms (2-3 worms) comes out of the cocoon. They are fully matured in 4 weeks.

Preservation

The worms are killed by dropping them in 70% ethyl alcohol. When the movement ceases, they are taken out from the alcohol and placed on a piece of blotting paper in a straight position. They are then transferred to a flat bottom container containing 10% formalin for fixation for at least overnight. The specimens are stored in suitable size vials filled with 70% alcohol & 5% formalin.

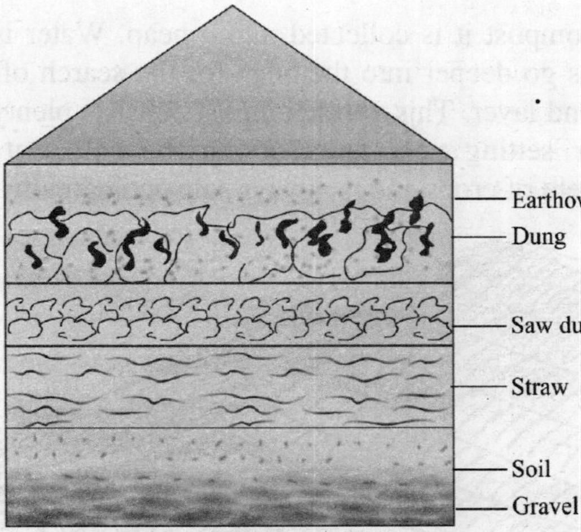

Fig. 1.7a: Vermiculture in Laboratory

Species suggested for Vermicomposting

In Indian conditions *Dichogaster bolani, Draivida willsi, Lampito mauritii* and *Perionyx excavatus* are best suited (Dash & Senapati, 1996).

African worm, *Eudrilus euginiae* has also been tried with promising results (Kale, 1986). *Eisenia foetida* and *Lumbricus rubellus* are commonly used species in temperate climate.

At Bharatpur, a NGO *Lupin Human Welfare & Research Foundation* is carrying out vermiculture at various villages with *Eisenia foetida*. 140 units are working in the district.

VERMICULTURE

Rearing of earthworms scientifically is called **Vermiculture.** The organic matter is decayed and transformed into manure called **Vermicompost.** It includes castings of earthworms, cocoons, undigested organic material and nutrients. Macro-nutrients like phosphorous, nitrogen, potassium and micro-nutrients like zinc, cobalt, copper and boron are found in vermicompost.

Vermicomposting can conveniently be done in 10 x 3 x 1.5 ft (length, width and height) size plot constructed on the ground under a hut shade. Temperature and humidity can be maintained. The said structure is filled with scrapes of crops like Bajra, sorghum, sugarcane, maize and rice upto 3" inch thickness. Dry compost or 4-5 days old dung is spread over it (2" inch thickness) and water is sprinkled to provide the moisture. A layer (1" inch) of vermicast (earthworms + cocoons + vermicompost) is spreaded over it. 25 kg of vermicast is sufficient for such structure. A further 2" inch thick layer of decayed dung and scrapes of vegetation, rotten, leaves and other debris is spreaded. A final layer of 10-12" inch thickness of dung and organic debris is spreaded and water to sprinkled to wet it. Now cover it with gunny bags soaked in water. Maintain the humidity by adding water time to time within 40-60 days, the debris is transformed into vermicompost. The organic matter is eaten by the worms and castings are accumulated into a heap. It is almost black. The earthworms are removed from the vermicompost for setting next unit. To remove earthworms from the vermicompost it is collected into a heap. Water is not sprinkled so that it begin to dry. The earthworms go deeper into the heap for the search of moisture and ultimately they are collected in the ground layer. This vermicompost consists plenty of worms and cocoons. It is used as vermicast for setting new units or can be sold out to other farmers. The vermicompost enhance the yield of crops and is always safe ecologically.

Fig. 1.7 (b) : Large scale vermiculture

Benefits of Vermicompost

1. Amount of humus is increased.
2. Microorganisms in the soil becomes active.
3. Nutrients helps in the increase in fertility.

Collection and Culture of Daphnia

Earlier stage of fish feeds on protein rich zooplanktons. Cladocerans are common zooplanktons found in fresh water bodies such as lakes, ponds, rivers, dams etc. *Daphnia carinata* is common cladoceran found in such sources. In Pisciculture, the mass culture of zooplanktons is very essential. It is done in circular tanks of 5,000 litre capacity. However, the laboratory culture can be done in acquarium or rectangular glass or acrylic tanks with aeration facilities. Cyclops and Daphnia can be collected from the natural water sources and cultured in small ponds or concrete tanks with stagnant water. After filling the tank with water it is manured with cow dung. When the water appears green due to the development of algae, it is inoculated with Daphnia collected from natural water sources. They grow rapidly and their density is reached 100 organism/ml within a week. They are harvested with the help of plankton net made up of bolting silk and given as feed to fish or shell fish larvae in the rearing tanks.

Identify the Daphnia and cyclops on the basis of the figures given below -

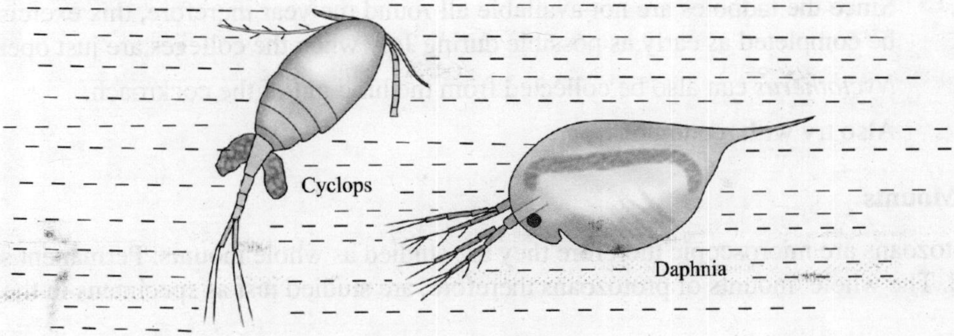

Fig. 1.8 : Cladocerans

Collection of rectal ciliates and other Intestinal Protozoa

Tadpoles of frogs and toads are best suited for the collection of rectal ciliates. Tadpoles can be collected from any pond, stagnant water temporary ponds in rainy season etc. during the month of July-August. Keep them in separate enamel finger bowls 1/2 filled with water.

Sacrifice the tadpole by destroying the brain with the help of needle (pithing). Cut and open the ventral wall from anterior to posterior end. Coiled intestine is visible. Posterior portion of the alimentary canal (rectum) is tightly coiled and separated from the intestine through a constriction. The rectal ciliates are distributed within the intestine and rectum. The anterior portion is inhabited by *Giardia* (a flagellate), *Nyctotherus* is commonly found in the rectum of tadpoles. *Opalina* is another ciliate found in the rectum of tadpoles.

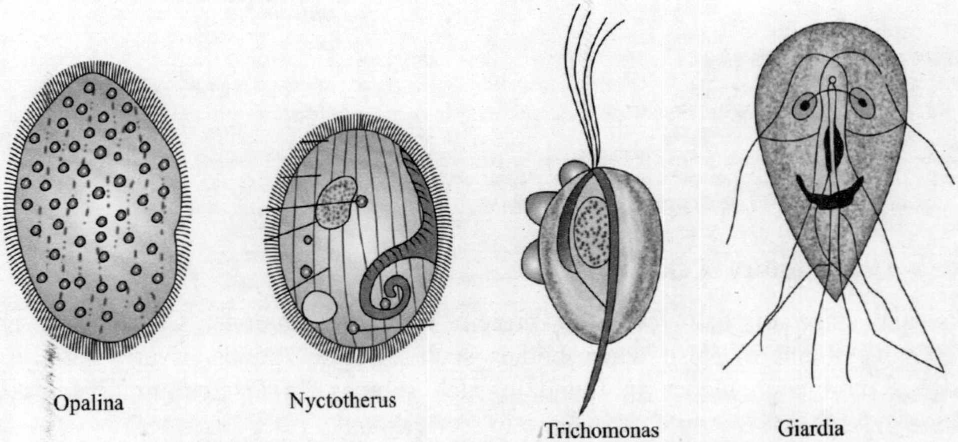

Fig. 1.8 (b). Nyctotherus, Opalina, Trichomonas, Giardia.

How to study in living condition?

Cut the part of the rectum on a clean slide in 0.7% saline solution. Tease the contents and cover them with cover slide. Identify even in the low magnification any of the species mentioned above.

Note

1. Since the tadpoles are not available all round the year therefore, this exercise should be completed as early as possible during July when the colleges are just opened.
2. *Nyctotherus* can also be collected from the hind gut of the cockroach.
3. Also try with rectum of frog.

Whole-Mounts

Protozoans are microscopic therefore they are studied as whole mounts. Permanent slides are prepared. The whole mounts of protozoans therefore, are studied just as specimens in the jar.

PROTOZOA (*Protos* - first, *Zooan* - animal)

Class

↓

SARCODINA	**CILIATA**	**FLAGELLATA**	**SPOROZOA**
↓ ORDER	↓ ORDER	↓ ORDER	↓ ORDER
1. **Lobosa** eg. Amoeba	1. **Holotricha** eg. Paramecium	1. **Dinoflagellata** eg. Ceratium, Noctiluca	1. **Gregarinida** eg. Monocystis
2. **Heliozoa** eg. Actinophrys	2. **Peritricha** eg. Vorticella	2. **Chrysomonadina** eg. Chrysamoeba	2. **Haemosporidia** eg. Plasmodium
3. **Radiolaria** eg. Acanthometra	3. **Chonotricha** eg. Spirochona	3. **Euglenoidina** eg. Euglena	3. **Coccidia** eg. Coccidium
4. **Foraminifera** eg. Polystomella	4. **Spirotricha** eg. Balantidium *Nyctotherus, Stentor*	4. **Volvocales/Phytomonadina** eg. Volvox, Chlamydomonas	4. **Cnidosporidia** eg. Myxobolus
	5. Opalinata eg. Opalina	5. **Choanoflagellate** eg. Proterospongia	5. **Sarcosporidina** eg. Sarcocystis
		6. **Protomonadina** eg. Trypanosoma	6. **Haplosporidia** eg. Haplosporidium
		7. **Hypermastigina** eg. Trichonympha	
		8. **Rhizo mastigina** eg. Mastigamoeba	
		9. **Polymastigina** eg. Trichomonas, Giardia	

Protozoa

1. Unicellular and microscopic animals.

Class : SARCODINA (RHIZOPODA)

1. Locomotory organs *Pseudopodia*.
2. Mode of nutrition is holozoic or saprozoic.

Order - Lobosa

1. Pseudopodia blunt

Order - Foraminifera

1. Shell multichambered
2. Reticulopodia emerge out from one or more openings.

Class - CILIATA

1. Locomotory organs *cilia.*

Order - Holotricha

1. All cilia of similar size

Order - Peritricha

1. Bell shaped body
2. Cilia arranged around oral region

Order - Spirotricha

1. Cilia encircling peristome, are larger arranged in clock wise direction.

Order - Opalinata

1. Cilia in oblique rows

Class - FLAGELLATA (MASTIGOPHORA)

1. Locomotory organs *Flagella.*

Order - Euglenoidina

1. Pellicle thick
2. Gullet leads into reservoir.

Order - Protomonadina

1. Parasitic in the blood
2. Flagella 2-8

Class - SPOROZOA

1. Locomotary organs absent
2. Endoparasitic

Order - Gregarinida

1. Extracellular trophozoite
2. Endoparasites

Order - Haemosporidia

1. Intracellular trophozoite
2. Blood parasite

EUGLENA

Classification	Phylum	-	Protozoa
	Class	-	Mastigophora
	Order	-	Euglenoidina
	Genus	-	Euglena

Common Name - Euglena

Habit & Habitat It is commonly found in fresh water ponds, ditches, Nutrition *holophytic*.

Distribution - Not specific, found almost everywhere in the said habitat.

Characters of Identification

1. Posterior end pointed and anterior end blunt.
2. Single flagella on the anterior end present which arise from the base of the gullet.
3. Stigma present near anterior end which is sensitive to light.
4. Nucleus near the posterior end, present.

Special Features

1. Chloroplasts radiating in the cytoplasm, present.
2. Reserve food in the form of *Paramylum* bodies.
3. They are called *organic puzzles* as they are half plant & half animals.

Animal like features

1. Contractile vacuole
2. Euglenoid movement
3. Holozoic nutrition

Plant like features

1. Presence of chloroplast
2. Holophytic nutrition
3. Palmella stage
4. Presence of pyrenoids

Biological Importance

1. Studied as representative protozoa in the laboratory.
2. Constitute planktons and is of great importance in aquatic ecosystem.

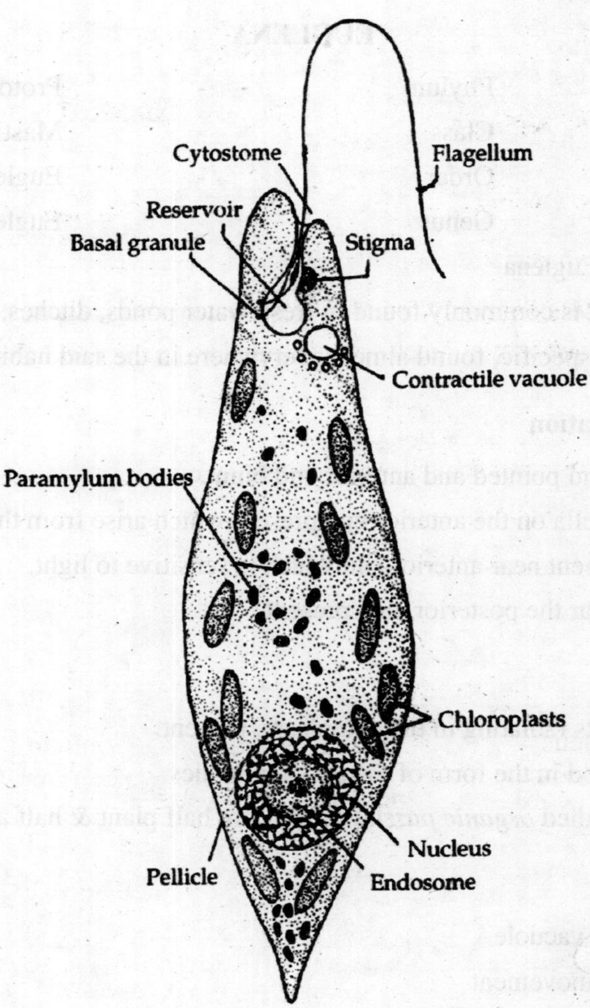

Fig. 1.9 : Euglena

AMOEBA

Phylum - Protozoa

Class - Rhizopoda (Locomotory organs - Pseudopodia)

Order - Lobosa (Pseudopodia blunt)

Genus - Amoeba (*Amoeba proteus*)

Common name : Amoeboid or no special name.

Habit and Habitat: Fresh water ponds and streams in decaying bottom vegetation.

Characters of Identification -

(1) Shape irregular and temporary cytoplasmic extensions called pseudopodia or false feet are found.

(2) Cytoplasm is differentiated into ectoplasm and endoplasm surrounded by plasma membrane.

(3) Endoplasm contain *food vacuole,* nucleus and contractile vacuole. It removes water from the amoeba.

Special characters

(1) Reproduction- asexual by *binary fission.*

(2) *Encystment* takes place in adverse environmental conditions.

(3) Amoeboid movement is considered the most primitive form of animal locomotion.

Economic importance

(1) Used in cell research for determining the relative functions and interactions of the nucleus and cytoplasm.

(2) *Entamoeba histolytica* causes amoebic dysentery

(3) *Entamoeba gingivalis* found in gingival pockets at the base of teeth and causes *pyrrhoea.*

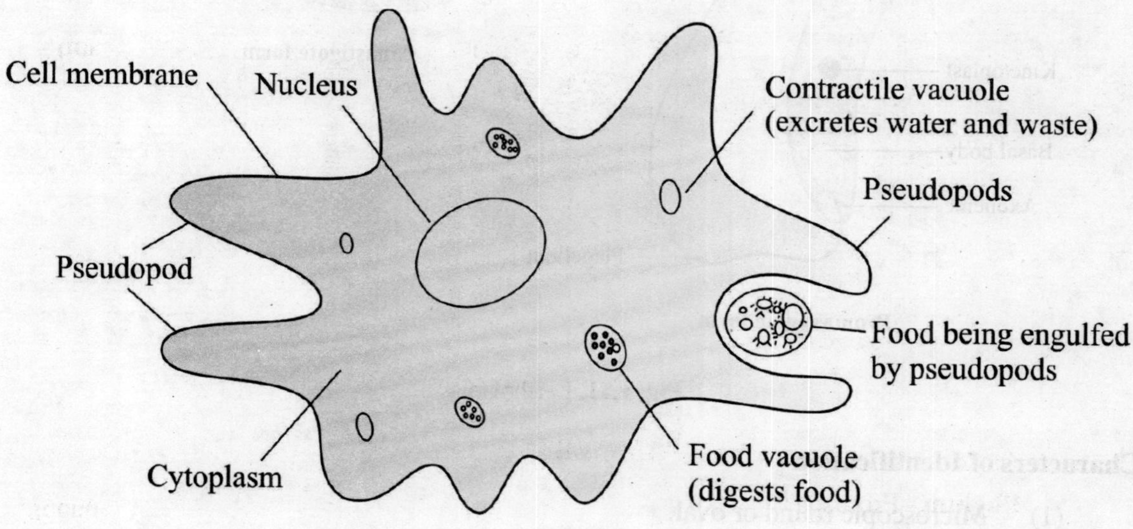

Fig. 1.10. Amoeba

LEISHMANIA

Phylum - Protozoa

Class - Mastigophora/Flagellata

Order - Protomonadina

Genus - Leishmania

(*Leishmania donovani*)

Common name - Leishmania (no special common name)

Habit & Habitat - Intracellular parasite in liver, intestine, spleen, lymph-glands and bone marrow. Pathogenic.

Distribution - Widely distributed in tropics and subtropics. In India, in Bihar, Orissa, Assam and West Bengal.

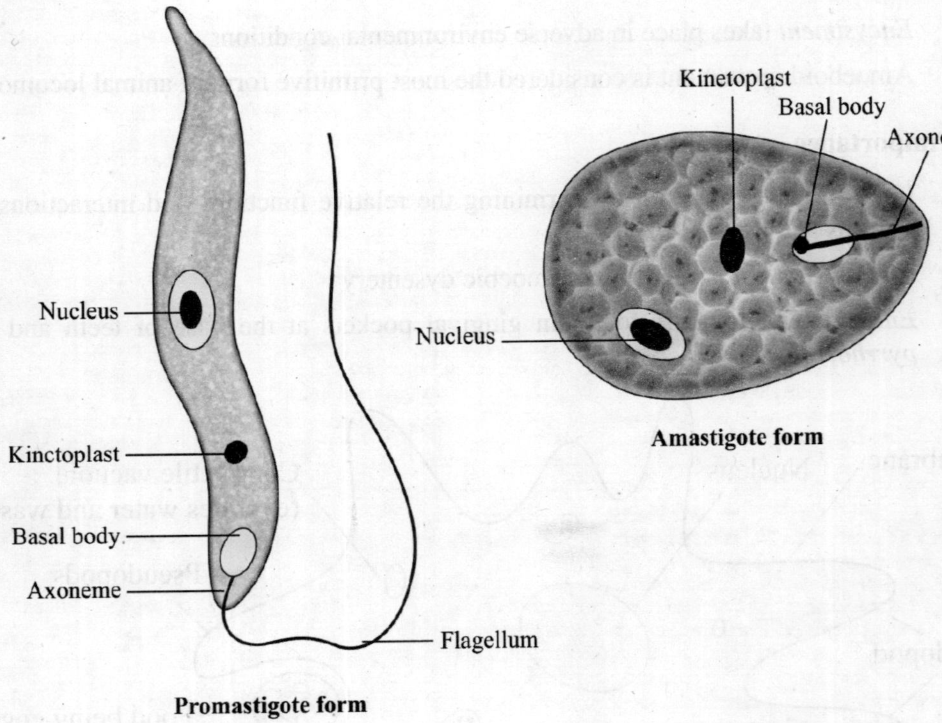

Fig. 1.11. Leishmania

Characters of Identification

(1) Microscopic round or oval.

(2) Body covered with pellicle (Periblast) cytoplasm is differentiated into ectoplasm and endoplasm.

(3) Exist in four forms (i) with no flagellum (ii) with flagellum and blepharoplast - anterior (iii) clithridia (subequatrial blepheroplast flagallum), (iv) with posterior blepheroplast - (trypanosoma)

Special characters

(1) Digenctic life cycle. Primary (Principal) host is vertebrate, secondary host - intermediate host or vector is **sand fly** (*Phlebotomus argentipes*).

Biological importance

(1) Leishmaniasis is linked to environmental changes such as deforestation, building of dams and urbanization.

(2) Disease is associated with malnutrition, poor housing, weak immune system and population displacement.

Economic importance

(1) *Leishmania donovani* causes disease *Leishmanasis/Kala azar.*

(2) *Leishmania tropica* causes *oriental-sore.*

There are 3 forms of Leishmaniasis

– *Visceral leishmaniasis* (Kala azar, most serious) characterised by fever, weight loss, enlargement of spleen and liver and anemia.

– *Cutaneous leishmaniasis* (most common, characterized by skin, lesion, ulcers)

– *Muco cutaneous leishmaniasis* - characterised by destruction of mucous membranes of nose, mouth and throat.

TRYPANOSOMA

Classification	Phylum	-	Protozoa
	Class	-	Flagellata
	Order	-	Protomonadina
	Genus	-	Trypanosoma
	Species	-	gambiens

Common Name - Trypanosoma

Habit and Habitat - They leads endoparasitic life in the blood and cerebro-spinal fluid (CSF) of man and alimentary canal of blood sucking invertebrates. Normal habitats are blood plasma, CSF, lymph nodes and spleen.

Distribution - Africa (Nigeria and Congo)

Characters of Identification

1. Slender elongated body with tapering ends.
2. Anterior end bears a single flagellum which arise from basal granule (blepharoplast)
3. Spherical nucleus in the middle of the body, present.
4. Body covered by thin pellicle which forms a fin like fold called *undulating membrane* laterally.
5. Reproduction by binary fission.

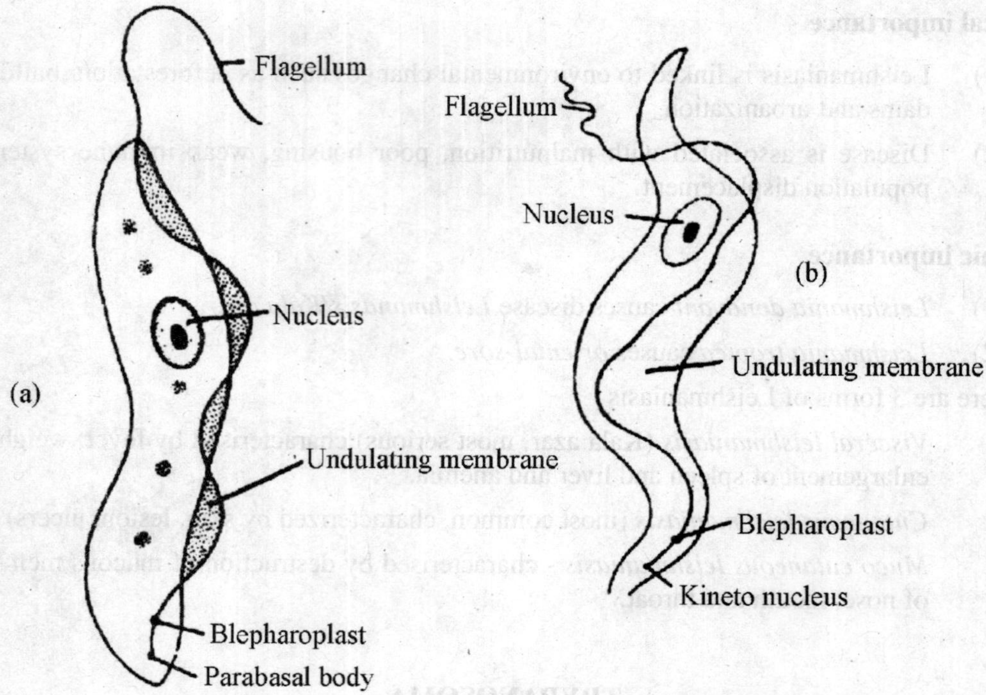

Fig. 1.12 (a) *Trypanosoma rhodesiense* (b) *Trypanosoma cruzi*

Special Features

1. It causes sleeping sickness or gambien **trypanosomiasis.**
2. The vector is *Glossina palpalis* (Tse-tse fly). It bites the man and feeds on the blood. (Secondary host).
3. Antelop, buffalo and pig are reservoir or temporary hosts. (Primary host)
4. In the intestine of Tse-tse fly, Trypanosomes reproduces. After two weeks they migrate to salivary glands from the gut. It is transmitted when the fly bites the man.
5. Sleeping sickness involves nervous system and lymphatic system. After incubation of 1-2 weeks. Fever, chills, headache and loss of appetite occur. Patient readily falls asleep at any time. At advance stage the patient goes in coma and ultimately dies.

Biological Importance

1. The whole mount is studied as representative protomonadina.

Economic Importance

1. It causes the disease *sleeping sickness* which affect the efficiency of man.

Additional Information

1. It was discovered by Forde in 1901.

2. **Sir David Bruce** reported that its transmission is through Tse-tse fly.
3. Drugs like Atoxyl, Bayer 205, Antripol, Germanin etc. are used in the control of the disease.
4. *Trypanosoma cruzi* spread **chaga's disease**, vector is a Reduvidae bug.

GIARDIA

Classification	Phylum	-	Protozoa
	Class	-	Flagellata
	Order	-	Polymastigina
	Genus	-	Giardia

Common Name - Giardia

Habit & Habitat - It is intestinal parasite in human beings.

Distribution - Almost all over the world

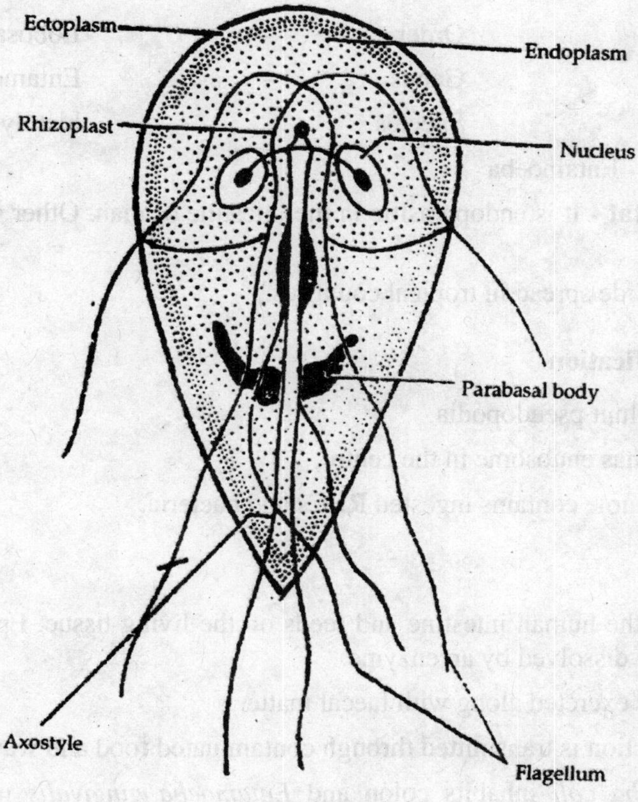

Fig. 1.13. Giardia

Characters of Identification

1. Bean shaped body with anterior end rounded and posterior pointed and tapering.
2. Ventral surface is concave and dorsal convex. The ventral surface possess a suctorial disc for the attachment of the host tissues.
3. Axostyle forms the median longitudinal axis.

Special Features

1. It occurs in the upper part of the small intestine and duodenum.

Biological Importance

1. It is studied in the Laboratory as representative polymastigina.

ENTAMOEBA

Classification	Phylum	-	Protozoa
	Class	-	Sarcodina
	Order	-	Lobosa
	Genus	-	Entamoeba
	Species	-	histolytica

Common Name - Entamoeba

Habit and Habitat - It is endoparasitic in the intestine of man. Other species inhabits colon and oral cavity.

Distribution - Wide spread in tropical countries.

Characters of Identification

1. It has a blunt pseudopodia.
2. Nucleus has endosome in the centre.
3. Food vacuole contains ingested RBCs and bacteria.

Special Features

1. It infest the human intestine and feeds on the living tissue. Epithelial linings of the colon are dissolved by an enzyme.
2. Cysts are excreted along with faecal matter.
3. The infection is transmitted through contaminated food and water.
4. *Entamoeba coli* inhabits colon and *Entamoeba gingivalis* in the oral cavity. *E. gingivalis* is transmitted through kissing.

Biological Importance

It is studied in the Laboratory as representative pathogenic protozoa.

Economic Importance

It causes *Amoebic dysentry (amoebiasis)*. *Entamoeba gingivalis* found in the mouth around the bases of teeth accelerate the disease *pyorrhea.*

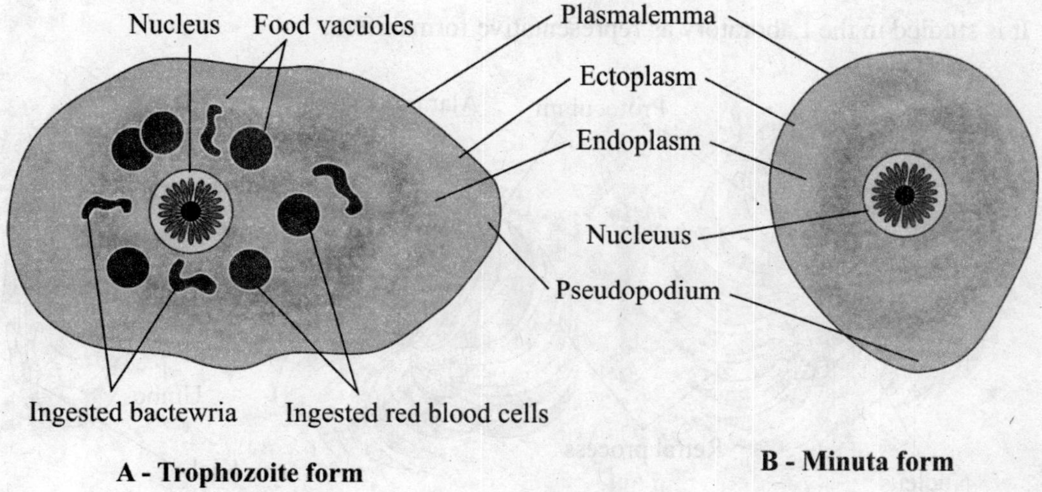

Fig. 1.14 (a) Entamoeba (b) Cyst

ELPHIDIUM (POLYSTOMELLA)

Classification	Phylum	-	Protozoa
	Class	-	Rhizopoda
	Order	-	Foraminifera
	Genus	-	Elphidium

Common Name - Polystomella

Habit and Habitat - It is marine found at the bottom of the sea. It feeds on diatoms and other protozoa.

Distribution - Not specified. Found in deep sea.

Characters of Identification

1. Calcareous perforated shell consisting of several chambers arranged spirally.
2. Pseudopodia projects through the perforations.

Special Features

1. It is *dimorphic, Microspheric* form possess small initial chamber whereas *megaspheric form* has large initial chamber.
2. Microspheric form reproduce sexually whereas megaspheric form by asexual means.
3. Dead shells constitute *foraminiferan ooze* at the bottom of the sea.

Biological Importance

It is studied in the Laboratory as representative formanifera.

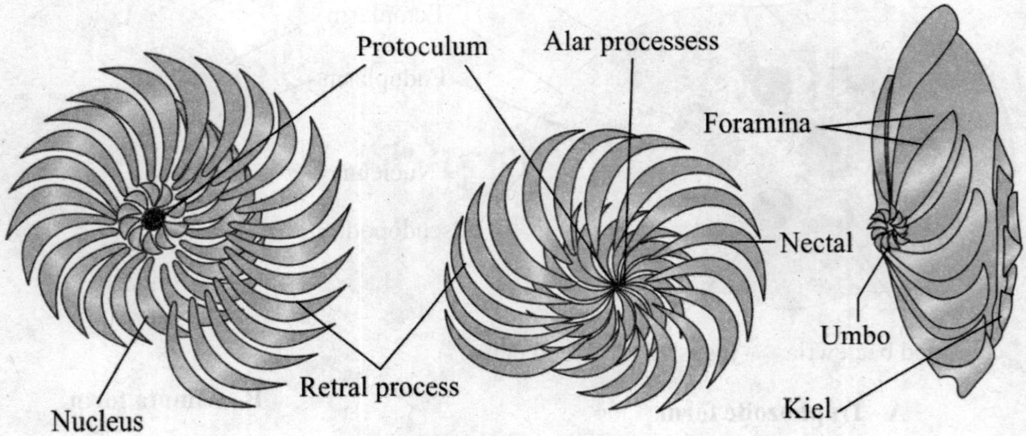

Fig. 1.15 : Polystomella

FORAMINIFERAN SHELLS

Classification	Phylum	-	Protozoa
	Class	-	Rhizopoda
	Order	-	Forminifera
	Genus	-	Foraminiferan shells

Habit and Habitat - Foraminiferans are found in sea. The empty shells lie on the sea bottom.

Distribution - Cosmopolitan.

Characters of Identification

1. Shells are unilocular and multilocular.
2. The shells of dead foraminera constitute **Foraminifera ooze.**
3. The shells are composed of *siliceous, chitinous* and *gelatinous* material.

Special Features

1. **Cambrian period** shells of foraminifera have undergone fossilization.
2. Limestone rocks of chalks in England, America (North) and France are exclusively foraminifera shells.

Biological Importance - They are studied in the Laboratory as representative foraminifera.

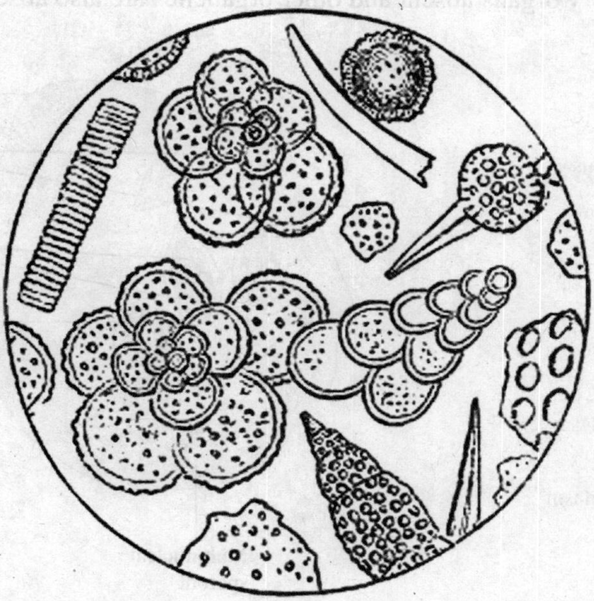

Fig. 1.16 Foraminifera Shells. (Rheophax, Globigerina, Nummulites, Pseudastrohiza, Cytlamnina, Turispirllina, Spiroloculina, Dendrophyra)

MONOCYSTIS

Classification	Phylum	-	Protozoa
	Class	-	Sporozoa
	Order	-	Gregarinida
	Genus	-	Monocystis

Common Name - Monocystis

Habit and Habitat

1. It is parasitic in the seminal vesicles of earthworm.
2. Saprozoic nutrition.

Distribution - Almost in all earthworms.

Characters of Identification

1. Trophozoite (feeding stage) is pointed at both the ends.
2. Single nucleus present
3. Pellicle present
4. Myonemes present
5. Locomotory organs absent and other organelles are also absent.

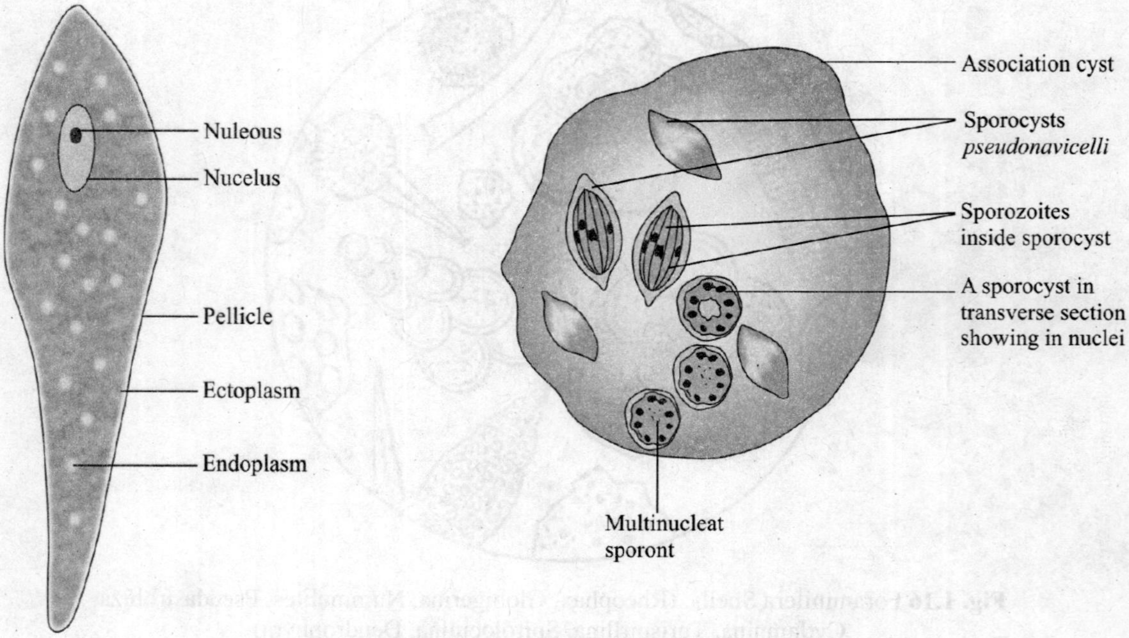

Fig. 1.17 (a) Monocystis (b) Sporozoites

Special Features

1. Trophozoite secrete proteolytic enzyme which dissolves the surrounding tissue which is absorbed by the general body surface.
2. They cause no harm to the host.
3. Boat shaped *pseudonavicella* infecting stage present.

Biological Importance

Studied in the Laboratory as representative gregarinida.

PLASMODIUM

Classification	Phylum	-	Protozoa
	Class	-	Sporozoa
	Order	-	Haemosporidia
	Genus	-	Plasmodium
	Species	-	vivax

Common Name - Commonly called malarial parasite.

Habit and Habitat - It is parasitic in the human RBCs (Intra-cellular parasite).

Distribution - Almost everywhere in India and other tropical countries of the world.

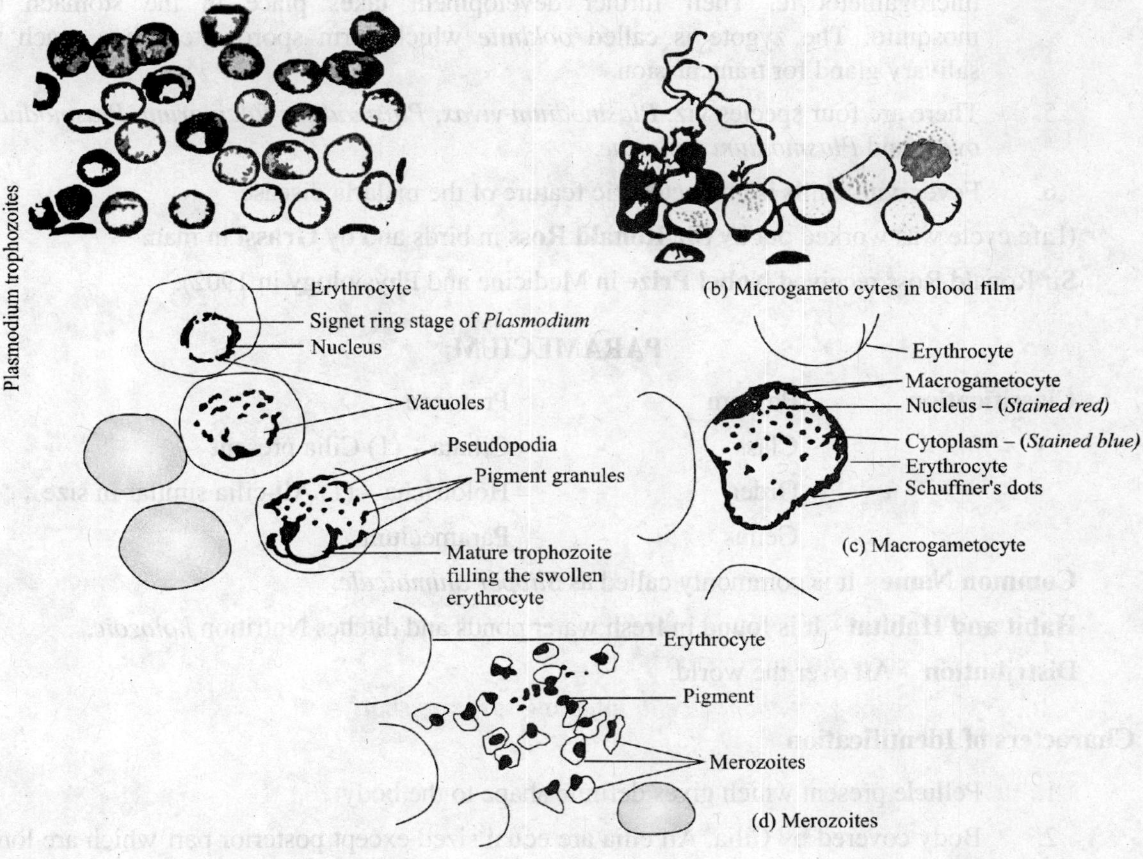

Fig. 1.18 : Different stages in life cycle of plasmodium

Characters of Identification

1. Sporozite is sickle shaped measuring 14 micron long and 1 micron broad.

2. It has no locomotory organs.
3. Single nucleus present.

Special Features

1. It causes *malaria fever*.
2. The life cycle is *digenetic* i.e. requires two hosts. Man is the *secondary host* whereas the mosquito is the *primary* host where sexual cycle is completed.
3. The infective stage sporozoite enters the human body through the bite of female *Anopheles* mosquito. Sporozoite reaches to liver cells and multiplies to form *merozoites.* Merozoites again multiplies by multiple fission through signet ring and amoeboid stages. This is called Schizogony.
4. After schizogony, some of the merozoites are modified into macrogametocyte and microgametocyte. Their further development takes place in the stomach of mosquito. The zygote is called *ookinite* which form sporozoite. They reach to salivary gland for transmission.
5. There are four species viz. *Plasmodium vivax, Plasmodium falciparum, Plasmodium ovale* and *Plasmodium malariae.*
6. Fever with chills is characteristic feature of the malaria disease.

(Life cycle was worked out by **Sir Ronald Ross** in birds and by **Grassi** in man.

Sir Ronald Ross received **Nobel Prize** in Medicine and Physiology in 1902).

PARAMECIUM

Classification	Phylum	-	Protozoa
	Class	-	Ciliata – (1) Cilia present
	Order	-	Holotricha – (1) All-cilia similar in size
	Genus	-	Paramecium

Common Name - It is commonly called as *Slipper animalcule.*

Habit and Habitat - It is found in fresh water ponds and ditches Nutrition *holozoic.*

Distribution - All over the world.

Characters of Identification

1. Pellicle present which gives definite shape to the body.
2. Body covered by Cilia. All cilia are equal sized except posterior part which are long called *Caudal tuft.*
3. Fixed contractile vacuole present with accessory vacuoles.
4. Cytoplasm is differentiated into ectoplasm and endoplasm.
5. Nucleus (macro and micro) present.
6. Feeding apparatus involves cytopharynx, cytostome, cytopyge and food vacuole.

Special Features

1. *Trichocysts* are present in the pellicle which are concerned with offense and defence.
2. Sexual reproduction by conjugation and asexual by binary fission.

Biological Importance

It is studied in the Laboratory as representative ciliate.

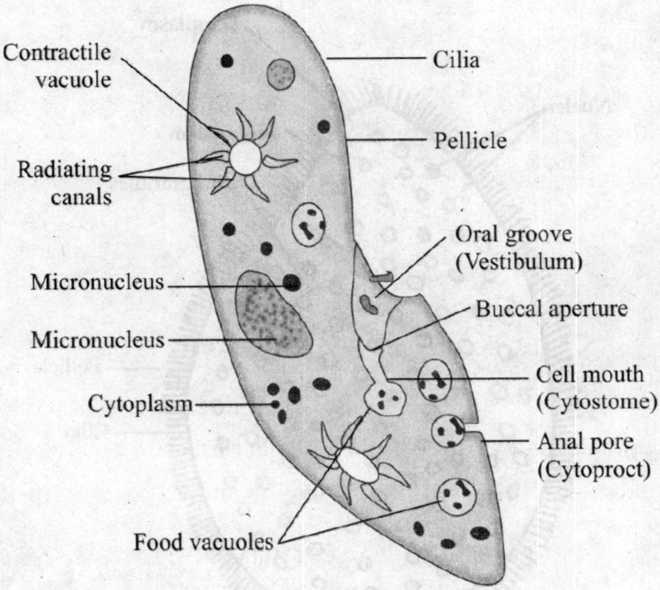

Fig. 1.19 : Paramecium

OPALINA

Classification	Phylum	-	Protozoa
	Class	-	Ciliata
	Order	-	Opalinata
	Genus	-	Opalina

Common Name - Opalina (Commonly known as rectal ciliates).

Habit and Habitat - Found in the rectum of tadpoles and adult frogs and toads they lead parasitic life.

Distribution - All places where frogs and toads are found.

Characters of Identification

1. It is large protozoa, oval in shape.
2. Cytoplasm differentiated into ectoplasm and endoplasm. Many nuclei present.
3. Myonemes and pellicle also present.

Special Features

1. It is Endo-commensel in the rectum of tadpoles.
2. Reproduction by binary fission and gamete formation.

Biological Importance

1. It is studied in the Laboratory as representative intestinal protozoa and ciliate.

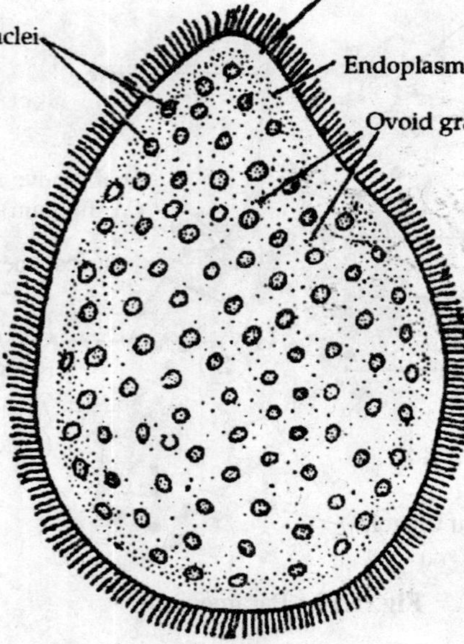

Fig. 1.20 : Opalina

NYCTOTHERUS

Classification Phylum - Protozoa
Class - Ciliata
Order - Heterotricha/Spirotricha (Cilia not of equal size)
Genus - Nyctotherus

Common Name - Nyctotherus

Habit and Habitat - Found in the intestine of frogs, toads, insects & mammals.

Distribution - All over the world.

Characters of Identification

1. Body bean shaped or kidney shaped.
2. Lateral peristomal groove present.

3. Contractile vacuole present.
4. Macro and micro nucleus present.
5. Body covered by cilia.

Special Features

1. Reproduction by binary fission and conjugation.
2. Endo-comensel.

Biological Importance

1. Studied in the Laboratory as representative intestinal protozoa.

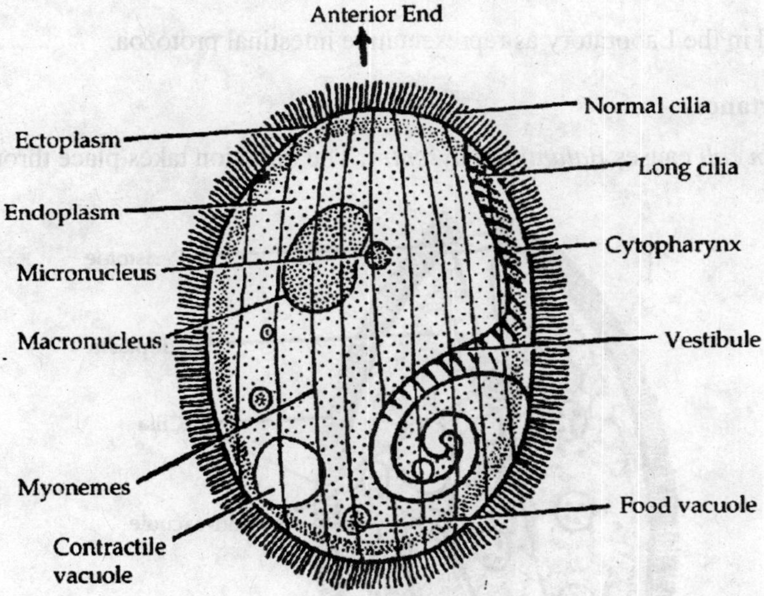

Fig. 1.21 : Nyctotherus

BALANTIDIUM

Classification	Phylum	-	Protozoa
	Class	-	Ciliata
	Order	-	Spirotricha
	Genus	-	Balantidium

Common Name - Balantidium

Habit and Habitat - Found in the intestine of man, monkey and pig and in the rectum of frogs and tadpoles.

Distribution - Cosmopolitan.

Characters of Identification

1. Appearance oval. Anterior end oval somewhat pointed.
2. Deep persistent groove present at the anterior end.
3. Body covered by cilia.
4. Large kidney shape nucleus and a micro-nucleus also present.
5 Contractile vacuole, Cytopharynx and Cytopyge present.

Special Features

1. Reproduction by conjugation, binary fission and encystment.

Biological Importance

It is studied in the Laboratory as representative intestinal protozoa.

Economic Importance

Balantidium coli causes *Balantidial dysentry*. The infection takes place through the cysts.

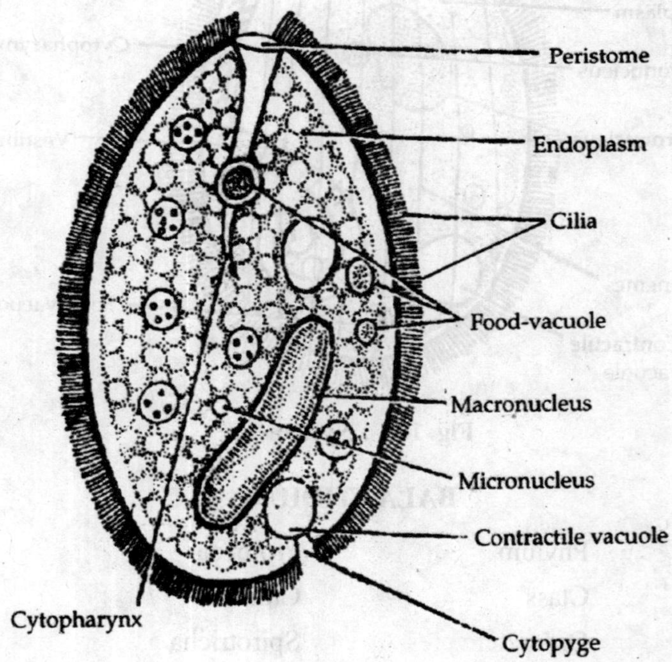

Fig. 1.22 : Balantidium

VORTICELLA

Classification - Phylum - Protozoa
Class - Ciliata

Order - Peritricha (Cilia present-surrounding the periostome)

Genus - Vorticella

Common Name - *Bell animalcule*

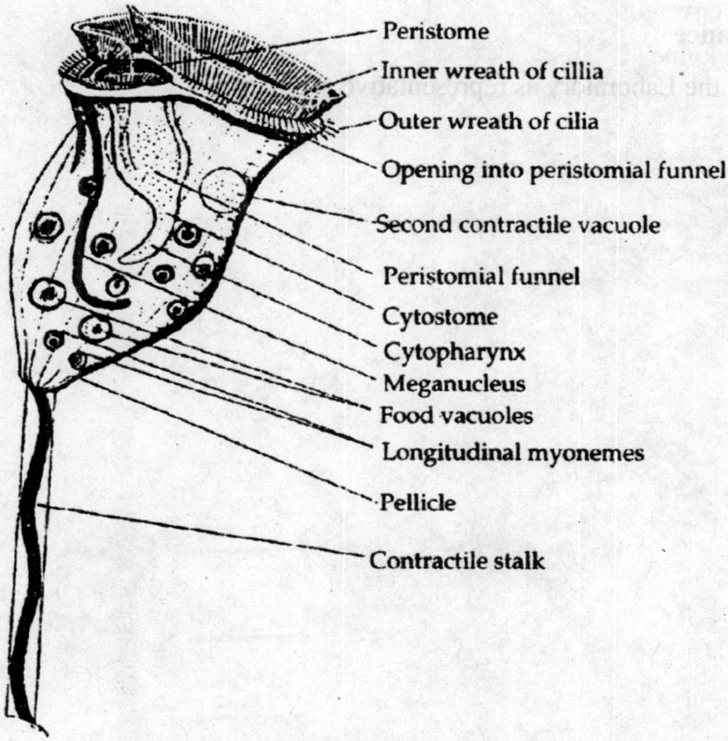

Fig. 1.23 : Vorticella

Habit & Habitat

1. Found in fresh water ponds attached to aquatic weeds.
2. Nutrition *holozoic*.
3. Solitary.

Distribution - Cosmopolitan

Characters of Identification

1. Bell shaped body. Pellicle present.
2. Peristomal groove leads into cytopharynx through the vestibule.
3. Contractile vacuole present.
4. Macro and micro nucleus present.
5. Two rows of cilia. The outer row encircle the rim of the bell and inner on the edge of the disc.

Special Features

1. Inverted bell like appearance.
2. Locomotion not found.

Biological Importance

It is studied in the Laboratory as representative protozoa.

Porifera (Poros - pores, ferre, - to bear)

Class

CALCAREA or CALCISPONGIAE	HEXACTINELLIDA or HYALOSPONGIAE	DEMOSPONGIAE
ORDER	ORDER	ORDER
1. **Homocoela** eg. Leucosolenia	1. **Hexasterophora** eg. Euplectella	1. **Tetractinellida** eg. Halisarca
2. **Heterocoela** eg. Sycon, Grantia, Scypha	2. **Amphidiscophora** eg. Hyalonema	2. **Monaxonida** eg. Spongilla
		3. **Keratosa/Ceratosa** eg. Euspongia & Hippospongiae
		4. **Myxospongiae** eg. Oscarella

Phylum - PORIFERA

1. Pore bearing animals
2. Cellular grade of organisation
3. Canal system present
4. Skeleton either made of *$CaCO_3$, spicules* or *silica* or horny *spongin fibres*
5. Sexual reproduction by gemmule formation

Class-CALCAREA

1. Skeleton is made up of $CaCO_3$ spicules

Order - Homocoela

1. Either Ascon, Sycon or Leucon type body and canal system

Order - Hetetrocoela

1. Solitary or colonial
2. Syconoid type canal system

Class-HEXACTINELLIDA

1. Marine
2. Skeleton is of six rayed siliceous spicules

Order - Hexasterophora

1. Hexaster spicules
2. Found attached to substratum

Class- DEMOSPONGIA

1. Skeleton of siliceous spicules or spongin fibres or both
2. Solitary or colonial

Order - Monaxonida

1. Monaxon spicules
2. Marine or fresh water

Order - Keratosa

1. Skeleton of spongin fibres

LEUCOSOLENIA

Classification -	Phylum	-	Porifera (Pore bearing animals)
	Class	-	Calcarea (Spicules of $CaCO_3$)
	Order	-	Homocoela (Spongocoel lined by choanocytes)
	Genus	-	Leucosolenia

Common Name - *Leucosolenia*

Habit and Habitat - It is found attached to stones, rocks, etc. in marine water.

Distribution - Cosmopolitan. Common in Europe.

Characters of Identification

1. It consist of vase shaped or upright cylinder which are united at the base.
2. Each cylinder represent a *olynthus stage* (simplest structures among sponges).
3. Spongocoel opens through osculum at the distal end.

Special Features

1. They are colonial.
2. Canal system is *ascon type*. The water current passes through the *ostia* into *spongocoel* and comes out through *osculum.*

   Water Current $\xrightarrow{\text{Ostia}}$ Spongocoel $\longrightarrow$ Osculum $\longrightarrow$ Out
3. Reproduction through budding, branching & regeneration.

Biological Importance

It is studied as representative Homocoela.

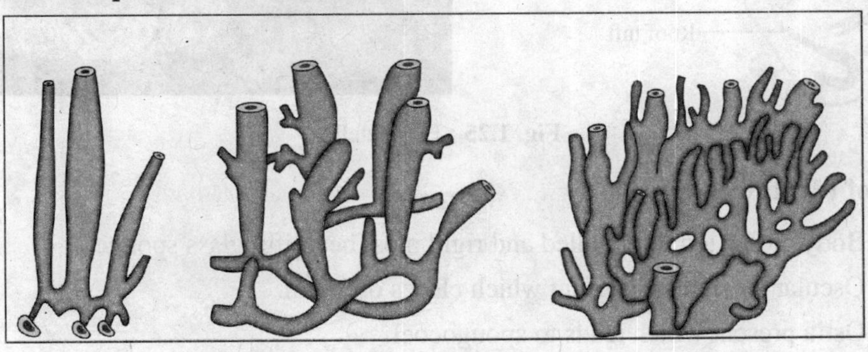

Fig. 1.24 : Leucosolenia (Progressive colony)

EUPLECTELLA

Classification - Phylum - Porifera

Class - Hexactinellida

(Skeleton of six rayed triaxon spicules)

Order - Hexasterophora

(Hexaster spicules present)

Genus - Euplectella

Common Name - *Venus Flower basket*

Habit and Habitat - Marine sponge inhabits deep waters (upto 5,000 meter)

Distribution - Found in the waters of West-Indies and Philippines.

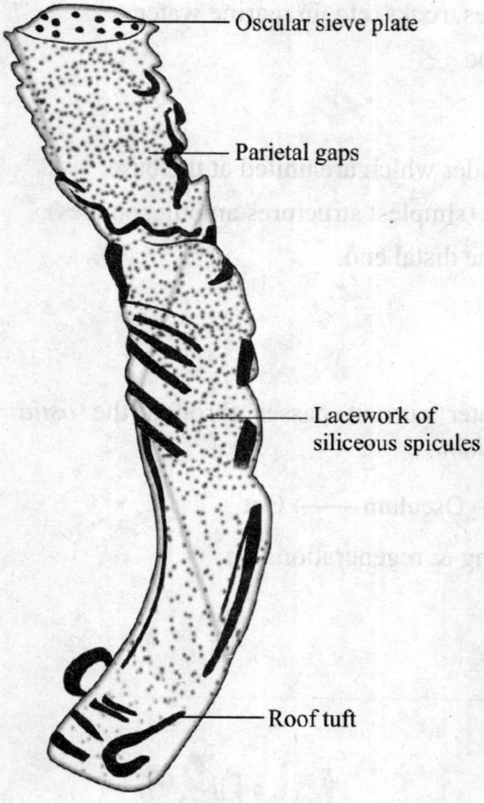

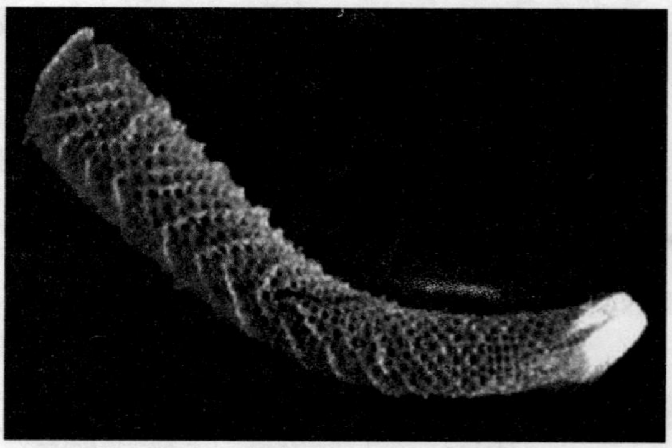

Fig. 1.25 : Euplectella

Characters of Identification

1. Body elongated thin walled and rigid most beautiful glass sponge.
2. Oscular sieve plate present which closes osculum.
3. Ostia present which leads to spongocoel.
4. Tuft of root spicules which helps in attachment at the substratum.

Special Features

1. In Japan it has earned a reputation of wedding gift because a crustacean (during copulation) gets entangled in it which hardly comes out of it. The crustacean is entangled for ever hence reflects a strong bond between bride and groom.

SPONGILLA

Classification - Phylum - Porifera

Class - Demospongiae

(1) Fresh water sponge

(2) Skeleton absent. If present, made up of spongin fibres or siliceous.

Order - Monaxonida

(1) Monoaxon spicules with or without spongin

Genus - Spongilla

Common Name - Commonly called *Fresh water sponge.*

Habit and Habitat - It is found attached to submerged stones, rocks, logs in fresh water lakes, ponds and streams.

Distribution - Pacific and Atlantic waters.

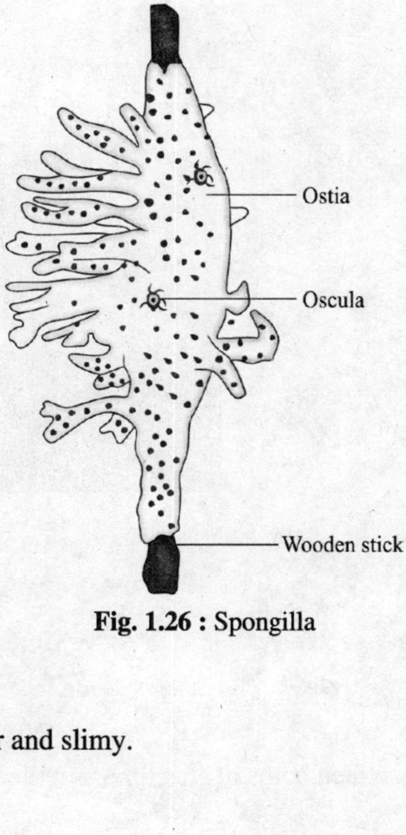

Fig. 1.26 : Spongilla

Characters of Identification

1. Colony is irregular and slimy.

2. Skeleton is made up of siliceous spicules embedded in spongin.
3. Colouration is yellowish brown or green.

Special Features

1. Gemmules (winter buds) are formed in winters. They sink to the bottom of the pond and rest till spring, afterward germinate into sponge.
2. Certain algae in its tissues, exhibit symbiotic relationship.
3. Canal system is *rhagon* type.

Biological Importance

1. It is studied in the Laboratory as representative Demospongiae

COELENTERATA / CNIDARIA (*Koilos - cavity, enteron-intestine*)

Class

HYDROZOA	**SCYPHOZOA**	**ANTHOZOA or ACTINOZOA**
ORDER	ORDER	ORDER
1. **Hydroidea** eg. Hydra	1. **Lucernaridae** eg. Lucernaria	1. **Stolonifera** eg. Tubipora Clavularia
2. **Hydrocorallina** eg. Millepora	2. **Cubomedusae** eg. Carydea	2. **Telestecea** eg. Telesto
3. **Trachylina** eg. Gonionemus	3. **Coranatae** eg. Periphylla	3. **Alcyonacea** eg. Alcyonium
4. **Siphonophora** eg. Physalia Vellela, Porpita	4. **Rhizostomeae** eg. Rhizostoma	4. **Coenolthecalia** eg. Heliopora
		5. **Gorgonacea** eg. Gorgonia, Corallium
		6. **Pennatulacea** eg. Pennatula, Virgularia, Renilla

Phylum - COELENTERATA

1. Diploblastic made up of layers viz. ectoderm and endoderm.
   A non-cellular *mesoglea* present between the two.
2. Tissue grade of organisation.
3. Coelenteron present.
4. Exhibit division of labour, colony dimorphic (polyp and medusae present)
5. Nematocyst as organs of offense and defence present.

Class - HYDROZOA

1. Polyps conspicuous.
2. Medusae stage either present or absent.

Order - Hydrocorallina

1. Calcacreous exoskeleton present.

Order - Siphonophora

1. Polymorphic

Class - SCYPHOZOA

1. Medusoid forms. Medusae umbrella like.

Order - Discomeduse (semaestomae)

1. Disc-shaped bell.

Class - ANTHOZOA

1. No medusoid form, only polyps present.

Order - Alcyonacea

1. Soft corals, spicules and polyps embedded in mesenchyme.

Order - Gorgonacea

1. Skeleton of *gorgonin* protein.
2. Tree like appearance

Order - Pennatulacea

1. Body divided into stalk and rachis with lateral branches.

OBELIA COLONY

Classification	Phylum	-	Coelenterata
			(1) Multi-cellular, triploblastic animals with radial symmetry.
	Class	-	Hydrozoa
			(1) Dimorphic or polymorphic
			(2) Alternation of generation found
	Order	-	Hydroidea
			(1) Sense organs from ectoderm
	Genus	-	Obelia

Common Name - *Obelia Colony*

Habit and Habitat- Colony is found attached to some sea weeds, stones, wooden logs. Marine.

Distribution - Found in shallow sea waters of Pacific ocean, Gulf of Mexico and California.

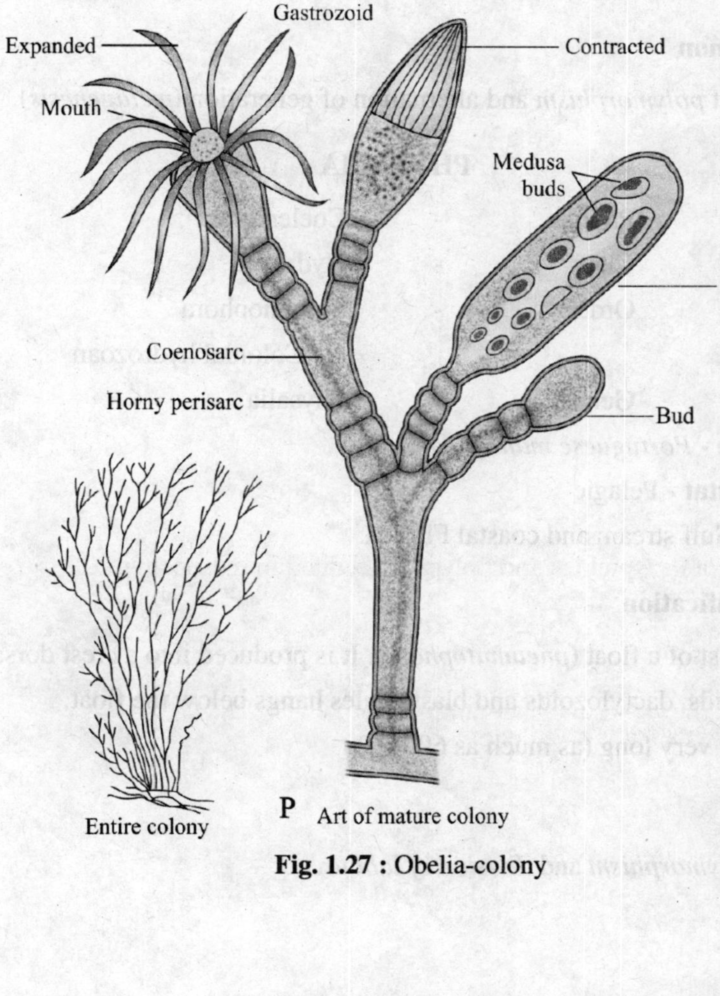

Fig. 1.27 : Obelia-colony

Characters of Identification

1. It is tree like colony with branches.
2. Colony consist of *hydrorhiza* (horizontal branch). Hydrorhiza give rise to *hydrocaulus* (upright branch) which end into zooids.
3. *Polyps* are present on the growing ends of the main branch, *blastostyle* at the base and *medusae* at lateral side of blastostyles.
4. Polyp is bell shaped with a mouth surrounded by tentacles. Blastostyle is without mouth and tentacles. Medusae saucer shaped.

Special Features

1. Obelia is a Polymorphic colony. Polyps serve vegetative, blastostyle nutritive and medusae reproductive function.
2. The tentacles have nematocyts for offence and defence purpose.

Biological Importance

It is studied in the Laboratory as a representative hydrozoan.

Biological Phenomenon

1. It exhibit *polymorphism* and alternation of generation (*metagenesis*).

PHYSALIA

Classification	Phylum	-	Coelenterata
	Class	-	Hydrozoa
	Order	-	Siphonophora
			(1) Colonial hydrozoan
	Genus	-	Physalia

Common Name - *Portuguese man of war*

Habit and Habitat - Pelagic

Distribution - Gulf stream and coastal Florida.

Characters of Identification

1. Colony consist of a float (*pneumatophore*). It is produced into a crest dorsally.
2. The gastrozoids, dactylozoids and blastostyles hangs below the float.
3. Tentacles are very long (as much as 60 feet)

Special Features

1. It exhibit *Polymorphism* and *division of labour*.

2. The tentacles bears powerful batteries of nematocysts which can kill a fish.
3. *Gastrozoid* helps in nutrition, *dactylozoids* are tactile and protective.

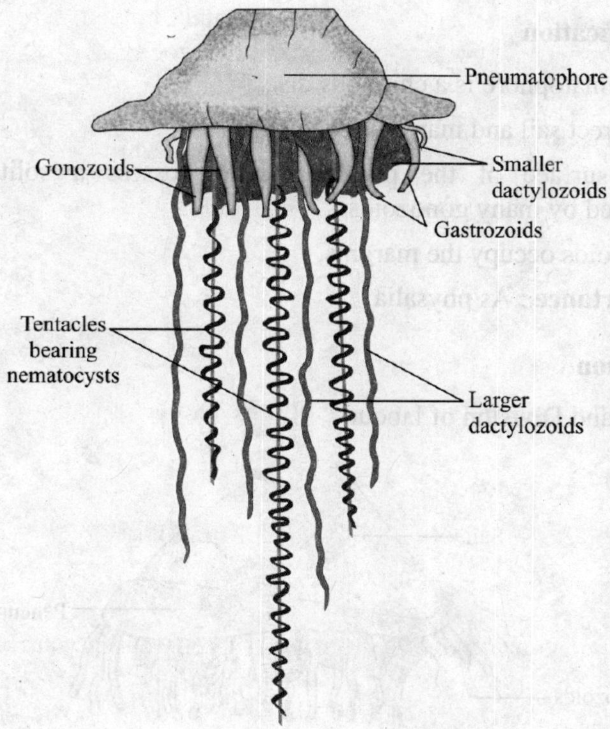

Fig. 1.28 : Physalia

Biological Importance

It is studied in the Laboratory as representative siphonophora.

Biological Phenomenon

Polymorphism and Division of labour.

Economic Importance

Sometimes inflict injuries to the swimmers as the tentacles bear powerful stinging cells.

VELELLA

Classification	Phylum	-	Coelenterata
	Class	-	Hydrozoa
	Order	-	Siphonophora
	Genus	-	Velella

Common Name - Commonly called *sail by wind.*

Habit and Habitat- Inhabits warm seas

Distribution - Atlantic coast (southern part), Pacific coast, Mexico.

Characters of Identification

1. The pneumatophore is a chitinous disc.
2. It has a erect sail and many air chambers.
3. Ventral surface of the pneumatophore possess a solitary central gastrozoid surrounded by many gonozoids.
4. Dactylozoids occupy the margins.

Biological Importance : As physalia

Biological Phenomenon

Polymorphism and Division of labour

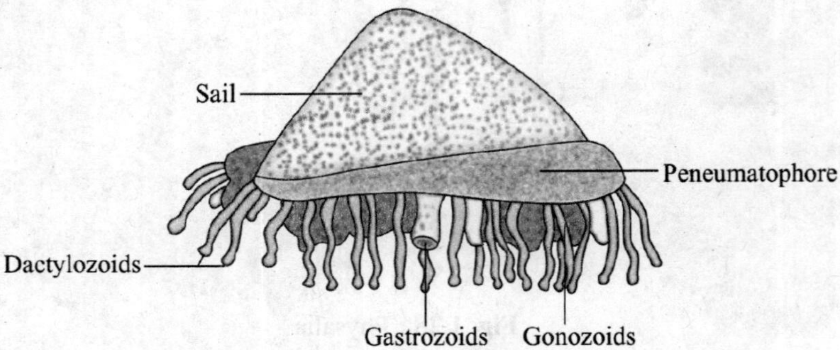

Fig. 1.29 : Velella

MILLEPORA

Classification	Phylum	-	Coelenterata
	Class	-	Hydrozoa
	Order	-	Hydroidea
			(1) Skeleton made up of $CaCO_3$
			(2) Colonial
	Genus	-	Millepora

Common Name - *Millepora*

Habit & Habitat- It is found in tropical marine shallow waters.

Distribution - Tropical seas.

Characters of Identification

1. It is a branching colony whose skeleton is made up of $CaCO_3$.

2. Numerous pores are located on the body.
3. Large pores are called *gastropores* & smaller one *dactylopores.*
4. *Gastrozoids* protrude from gastropores and *dactylozoids* from dactylopores.

Special Features

1. Dactylopores serves the function of protection and tactile also.

Biological Importance

It is studied in the Laboratory as representative Hydrocoralina.

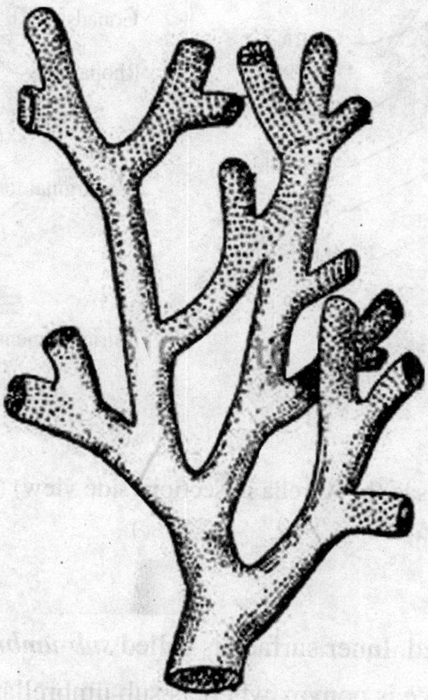

Fig. 1.30 : Millepora

AURELIA

Classification	Phylum	-	Coelenterata
	Class	-	Scyphozoa
			(1) Medusae stage conspicuous
			(2) True velum absent
	Order	-	Discomedusae/Semaestomea
			(i) Disc shaped
			(ii) Tentaculocysts 8 or > 8
	Genus	-	Aurelia

Common Name - It is commonly called *Jelly fish.*

Habit and Habitat- Found in marine coastal waters near the surface of the sea.

Distribution - - Pacific ocean.

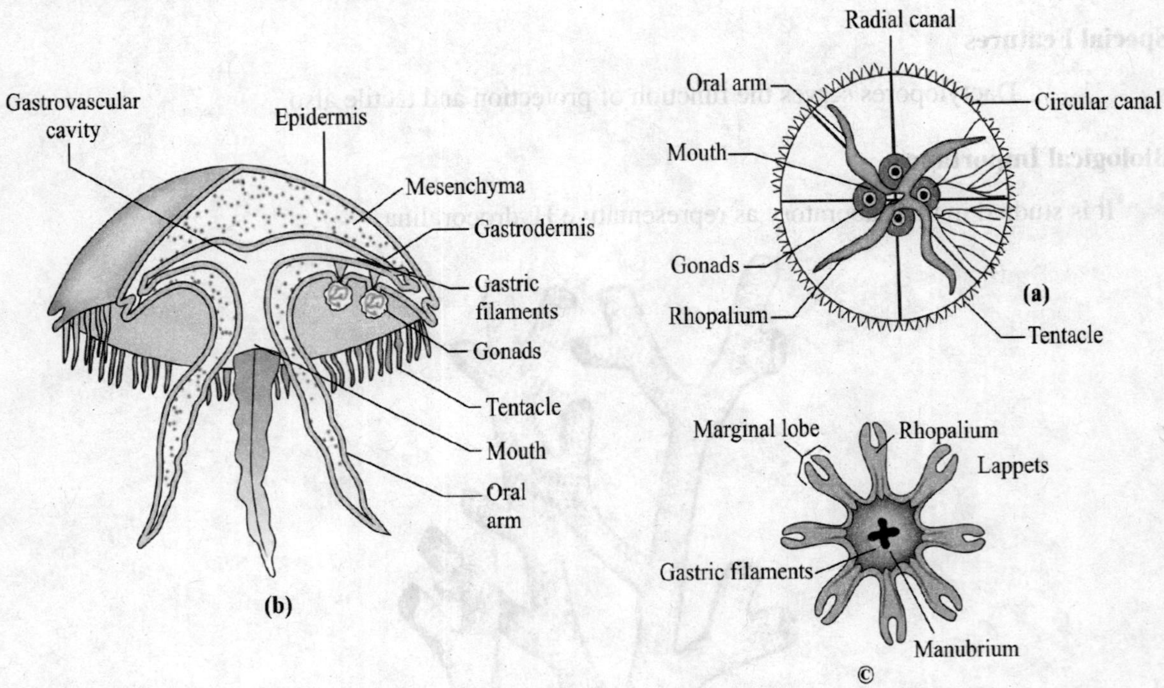

Fig. 1.31 (a) Aurelia medusae (b) Aurelia in section (side view) (c) Ephyra larva of aurelia

Characters of Identification

1. It is umbrella shaped. Inner surface is called *sub-umbrellar,* outer is *ex umbrellar*.
2. Ex-umbrellar surface is convex whereas sub umbrellar is concave.
3. Manubrium bears a mouth.
4. From the mouth arise hanging four oral arms.
5. Sub genital pits are located near the mouth.
6. Red coloured gonads are situated near the sub-genital pits.

Special Features

1. Larva is *ephyra*.
2. It exhibit *strobilization* (Scyphistoma divides by transverse fission)

Biological Importance

1. It is studied in the Laboratory as representative scyphozoa.

ALCYONIUM

Classification	Phylum	-	Coelenterata
	Class	-	Anthozoa
	Order	-	Alcyonaria (Lower part of the polyps fused)
	Genus	-	Alcyonium

Common Name - *Dead man's finger*

Habit and Habitat- Found attached to stones in marine water below low tide mark.

Distribution - Temperate and cold sea.

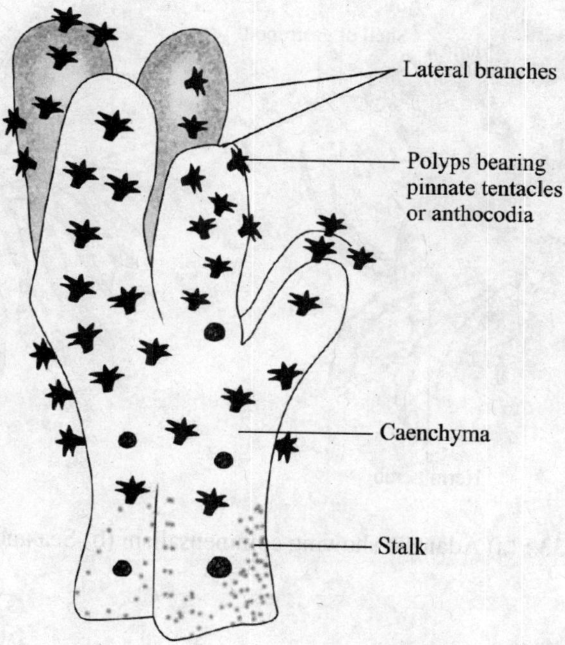

Fig. 1.32 : Alcyonium

Characters of Identification

1. Colony is branched into lobes or digits.
2. Numerous polyps protrude out from the pores.
3. Skeleton is made up of calcareous spicules embedded in mesoglea.

Special Features

1. Each polyp consist of *basal part* and a distal called *anthocodium.*
2. Anthocodium has a mouth and eight pinnate tentacles.

Biological Importance

1. It is studied in the Laboratory as a representative alcyonaria.

SEA ANEMONE

Classification

Phylum	-	Coelenterata
Class	-	Anthozoa
Order	-	Actinaria (1) Skeleton absent
Genus	-	Adamsia

Common Name - *Sea anemone* itself is a common name of *Adamsia*, *Metridium* is also called sea anemone.

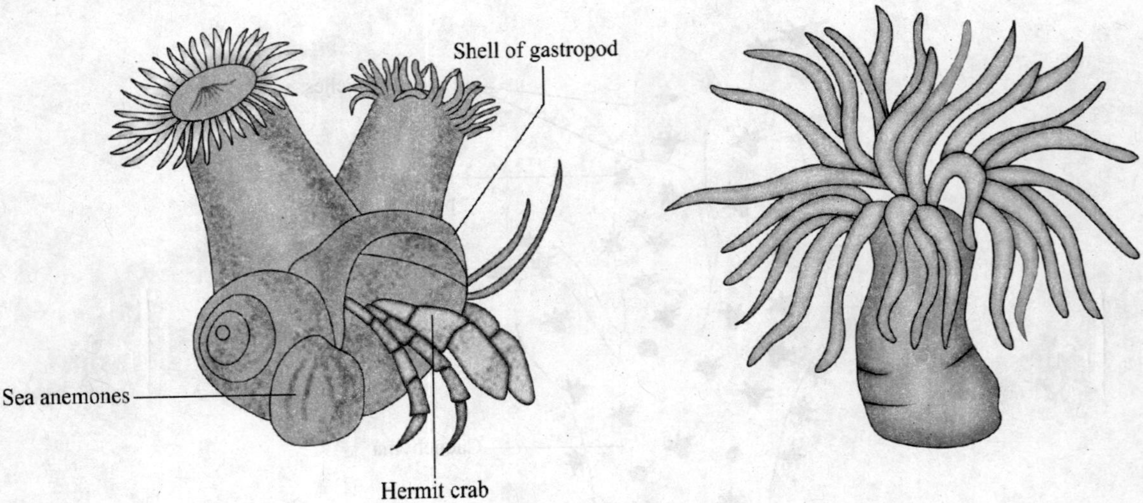

Fig. 1.33 : (a) Adamsia showing commensalism (b) Sea anemone

Habit and Habitat

1. Found attached to wood, stones, sea weeds and shells in marine waters.
2. Solitary, sessile.

Distribution - Pacific oceans, Atlantic coast and Europe.

Characters of Identification

1. Lower part of the body constitute *basal disc* which attach to the substratum. It is highly muscular and glandular.
2. Upright part of the body is called *column*.
3. *Oral disc* is surrounded by tentacles, bears a slit like mouth.

Special Features

1. Some anemone show *symbiotic* relationship with other animals, e.g. Gastropod shell and hermit crab (*commensalism*).

Biological Importance

It is studied in the Laboratory as representative Anthozoa.

GORGONIA

Classification	Phylum	-	Coelenterata
	Class	-	Anthozoa (Actinozoa)
	Order	-	Gorgonacea
			(i) Tree like colonies
			(ii) Skeleton of gorgonin
	Genus	-	Gorgonia

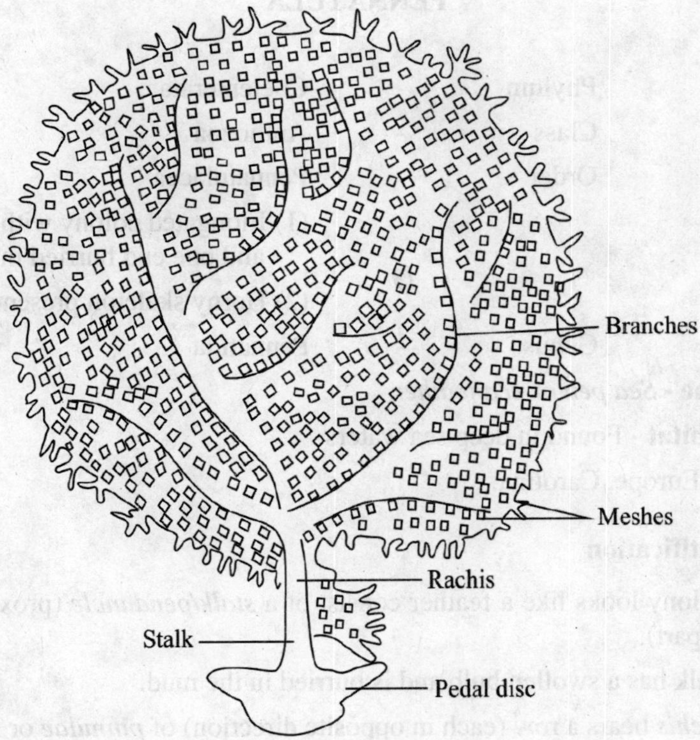

Fig. 1.34: Gorgonia

Common Name - *Sea fan*
Habit and Habitat - Found in warm marine waters.
Distribution - Atlantic ocean

Characters of Identification

1. The colony is tree like branching.
2. The branches arise from the main trunk branches fastened to the substratum by basal plate.

Special Features

1. The skeleton contains horn like proteinaceous material called *gorgonin*.
2. It constitute a major part of the ``under water gardens". They are of various shades & colouration.

Biological Importance

It is studied in the Laboratory as representative Gorgonacea.

PENNATULA

Classification	Phylum	-	Coelenterata
	Class	-	Anthozoa
	Order	-	Pennatulacea
			(1) Elongated colony with polyps free and one end burried in mud.
			(2) Horny skeleton present.
	Genus	-	Pennatula

Common Name - *Sea pen or sea feather*

Habit and Habitat - Found in deep sea waters.

Distribution - Europe, Carolina.

Characters of Identification

1. The colony looks like a feather consist of a *stalk/penduncle* (proximal) and a *rachis* (distal part).
2. The stalk has a swollen bulb and is burried in the mud.
3. The *rachis* bears a row (each in opposite direction) of *pinnulae* or leaves.

Special Features

1. The colony is *dimorphic*. The *autozoids* occur in single row on the upper end of pinnulae and small siphonozoids are confined to the dorsal side of the rachis.
2. Skeleton consist of calcareous spicules.

Biological Importance

It is studied in the Laboratory as representative Pennatulacea.

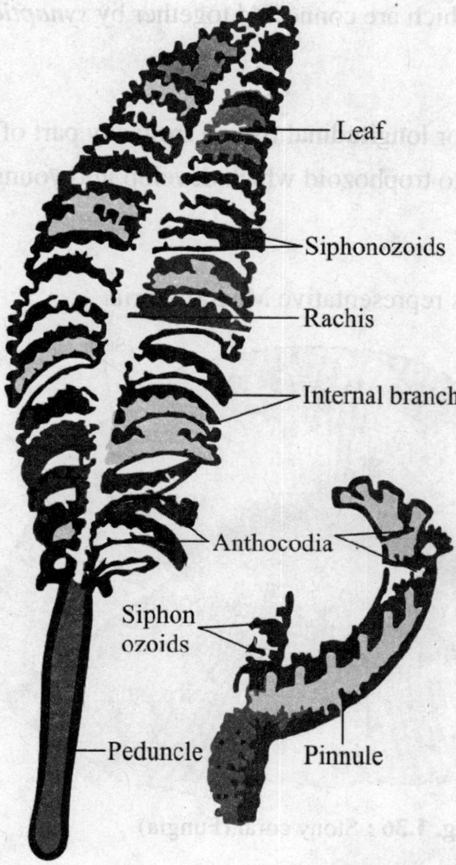

Fig. 1.35 : Penntaula

FUNGIA (STONY CORAL)

Classification	Phylum	-	Coelenterata
	Class	-	Anthozoa
	Order	-	Madreporaria
			(1) Colonial called stony coral
			(2) Skeleton made up of calcareous rods
	Genus	-	Fungia

Common Name - Commonly called *Mushroom coral.*

Habit and Habitat - Found at the bottom of the sea.

Distribution - Gulf of California.

Characters of Identification

1. Lower side concave and upper is convex.
2. Body has many septa which are connected together by *synapticulae*.

Special Features

1. Reproduce by budding or longitudinal fission from any part of the corallite.
2. *Planula* larva give rise to trophozoid which develop into young coral.

Biological Importance

It is studied in the Laboratory as representative Madreporaria.

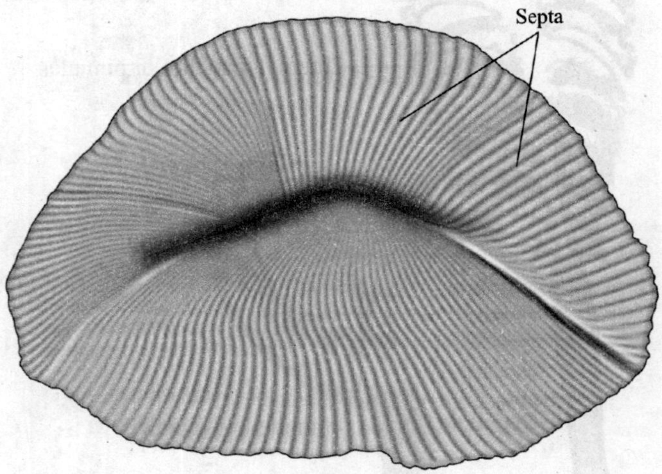

Fig. 1.36 : Stony coral (Fungia)

CTENOPHORA/ACNIDARIA (*Ctenes - Comb, Ferre to* bear)

TENTACULATA	**NUDA**
(ORDER)	(ORDER)
1.**Cydippida** eg. Pleurobranchia	1. **Beroida** eg. Beroe
2. **Lobata** eg. Deiopea	
3. **Cestida** eg. Cestum	
4. **Platyctenea** eg. Coeloplana, Ctenoplana	

Phylum - CTENOPHORA

1. Marine, Tissue grade and bi-radial symmetry animals.
2. Stinging cells nematocysts absent. Colloblast for adhession present. Locomotion through comb-rows.
3. Circulatory, respiratory systems absent.
4. Pelagic

Class - Tentaculata

1. Tentacles present

Order - Cydippida

1. To long opposite tentacles present

Class - NUDA

1. Tentacles absent

Order - Beroida

1. Laterally compressed body.

PLEUROBRANCHIA

Classification Phylum - Ctenophora
(i) Bi-radial symmetry
(ii) Mouth on one end & sense organ on the other
Class - Tentaculata
(i) Two tentacles present
Order - Cydippida
(i) Cydippid larva
Genus - Pleurobranchia

Common Name - *Comb jelly* or *Sea Walnut* or *Sea gooseberry*

Habit and Habitat - Marine, pelagic and solitary. Feeds on planktons.

Distribution - Pacific coast.

Characters of Identification

1. Body transparent and spherical.
2. Mouth is situated at oral pole whereas sense organs on the aboral pole.
3. Biradially symmetrical. Eight equally spaced comb plates arranged on the side of the body. The comb rows are composed of a short ciliary plate or ctenes.
4. Two long solid retractile tentacles present in opposite side of the body each project from a tentacle sheath.

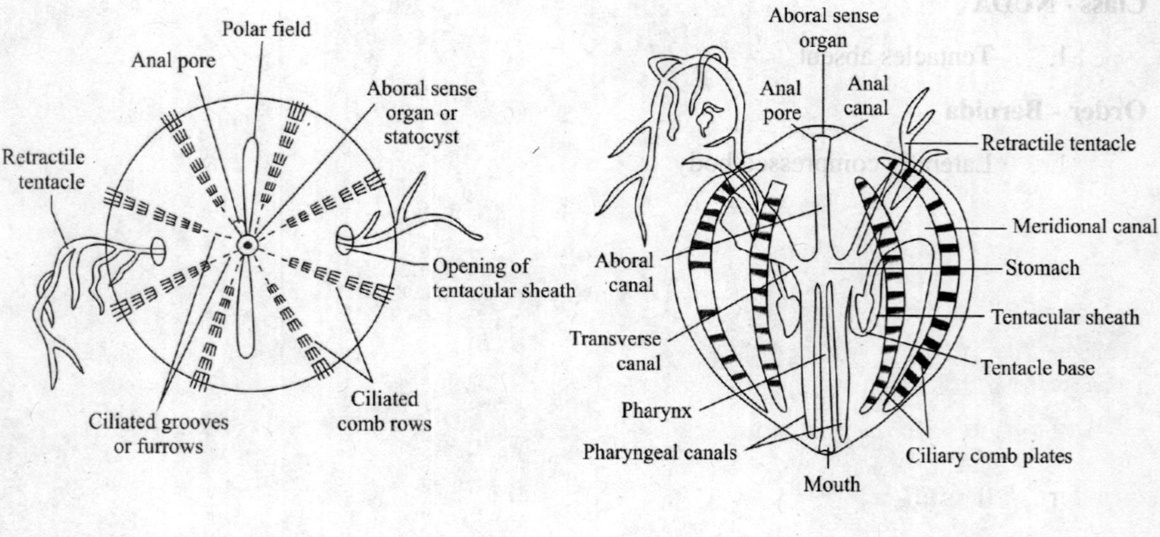

Fig. 1.37 : (a) Pleurobranchia (Aboral view) (b) Pleurobranchia (side view)

Special Features

1. Hermaphrodite.
2. Statocyst as a sensory organ.
3. Some ctenophore exhibit *neoteny* (both larva and adult reproduce sexually).
4. They possess great power of *regeneration*.
5. They exhibit **luminescence.**
6. They show affinites with *Cnidaria* and *platyhelminthes*. They are intermediate between Radiata and Bilateria.

Cnidarian Characters

1. Biradial symmetry.
2. Diploblastic body.
3. Mesoglea present.
4. Coelom absent.

Platyhelminthes like Characters

1. Dorsoventrally flattened.
2. Ciliated epiderms.
3. Crawling mode of locomotion.

Biological Importance

1. It is studied in the Laborotory as representative Ctenophore.

BEROE

Classification	Phylum	-	Ctenophora
	Class	-	Nuda
	Order	-	Beroida
	Genus	-	Beroe

Common Name - Beroe.

Habit & Habitat - It is voracious feeder, feeds on crustaceans. Found in cold sea waters.

Distribution - Mediterranean sea and Pacific oceans.

Characters of Identification

1. Conical body with aboral and oral surfaces.
2. Sense organs on aboral surface and mouth on oral surface.

3. Pharynx enlarged.
4. 4 comb plates present.

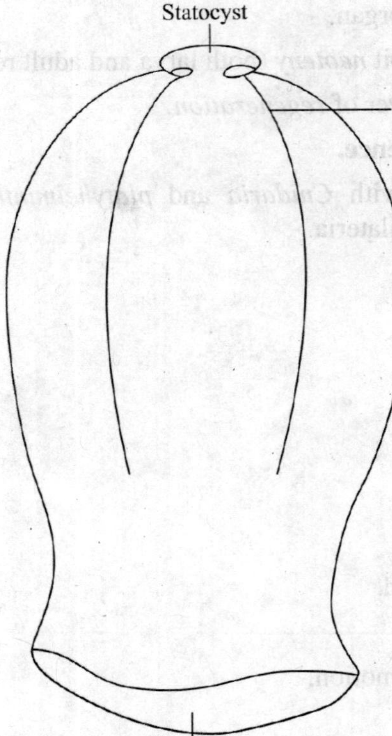

Fig. 1.38 : Beroe

Special Features

1. Hermaphrodite.
2. Exhibit *luminescence*, Light production takes place in the walls of meridional canals.
3. Gastrula develops into free swimming ***cydippid larva***.

Biological Importance

It is studied in the Laboratory as representative Ctenophore.

Outline classification of Platyhelminthese

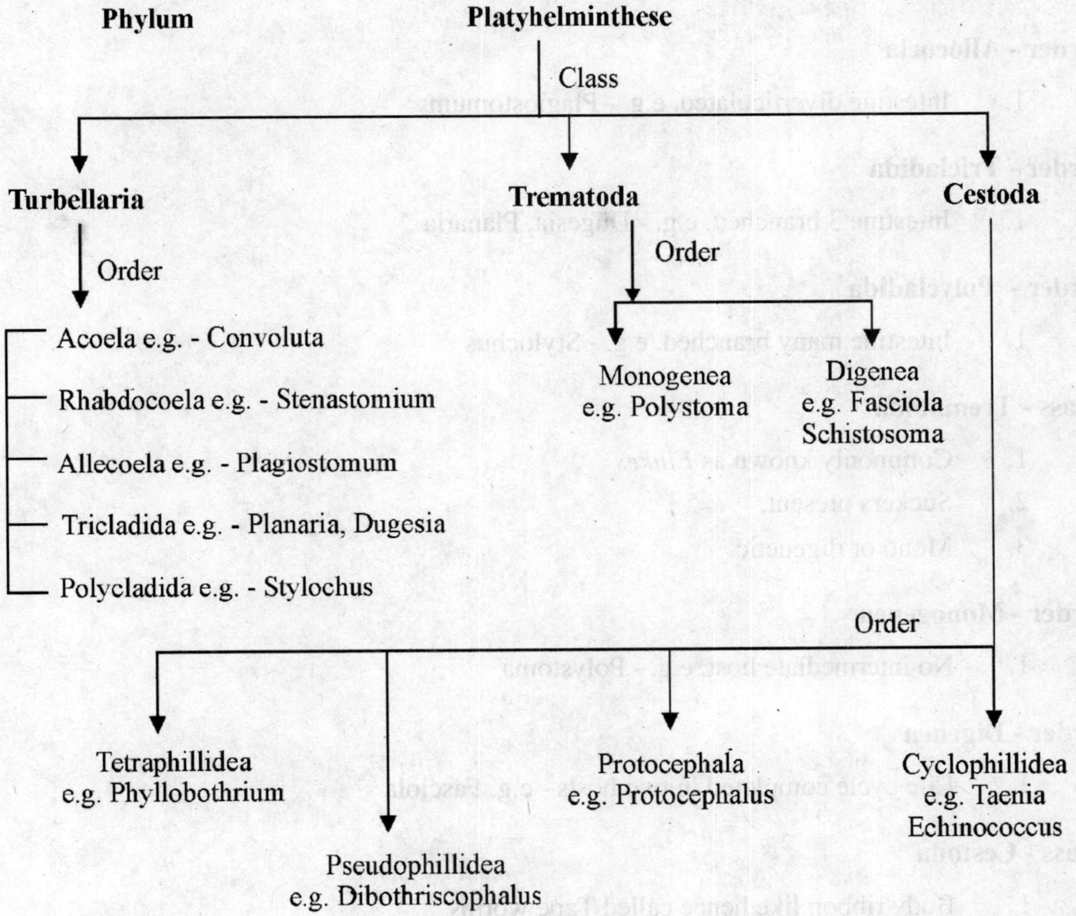

Platyhelminthese

1. Commonly called *flat worms*.
2. Triploblastic, Acoelmate with organ grade of organisation.
3. Free swimmng or parasitic.

Class - Turbellaria

1. Free swimming

Order - Acoela

1. Marine
2. Intestine wanting. e.g. - Convoluta.

Order - Rhabdocoela

1. Tricastine sac like. e.g. - Stenostomum

Order - Allococla

1. Intestine diverticulated. e.g. - Plagiostomum

Order - Tricladida

1. Intestine 3 branched. e.g. - Dugesia, Planaria

Order - Polycladida

1. Intestine many branched. e.g. - Stylochus

Class - Trematoda

1. Commonly known as *Flukes*
2. Suckers present.
3. Mono or digenetic

Order - Monogenea

1. No intermediate host. e.g. - Polystoma

Order - Digenea

1. Life cycle completed in two hosts - e.g. Fasciola

Class - Cestoda

1. Body ribbon like hence called Tape worms
2. Endoparasitic
3. No intestine present.

Order - Tetraphillidea

1. Scolex with 4 bothridia
2. Parasites of elasmobranchs. e.g. - Phyllobothrium

Order - Pseudophillidea

1. Scolex with 2 bothridia. e.g. - Dibothricephalum

Order - Cyclophillidea/Taenioidea

1. True suckers present.
2. Endoparasites of mammals and birds. e.g. - Taenia, Echinococcus.

Planaria (Whole mount)

Classification

Phylum - Platyhelminthese

Class - Turbellaria (1) Free Swimming, non segmented worm.

Order - Tricladida (1) Alimentary canal 3 branched.

Genus - Planaria

Common name : Planaria, Dugesia is also called Euplanaria.

Habit & Habitat

It is free swimming in fresh water or Pond, river or lakes found attached to some stones or log or mud at bank. It is carnivorous.

Distribution

Temperate regions, India, Britain and America.

Characters of Identification

1. Body leaf like, slightly broad at anterior end.
2. Head triangular that bears a pair of eyes.
3. Lateral projections of head called *auricles.*
4. Dorso-lateral side and ventral surface is covered with cilia.
5. Pharynx is protrusible that opens on the ventral surface.
6. Genital pore on middle part.

Special characters

1. Power of *regeneration* enormous
2. Alimentary canal 3 branched (tricladida)
3. Excretion through *flame cells*
4. *Photo negative.*

Biological Importance

1. Studied as Laboratory animal as microscopic slide (w.m.).
2. Used as research animal in ethology on phototactic response.
3. Used as research animal on regeneration.

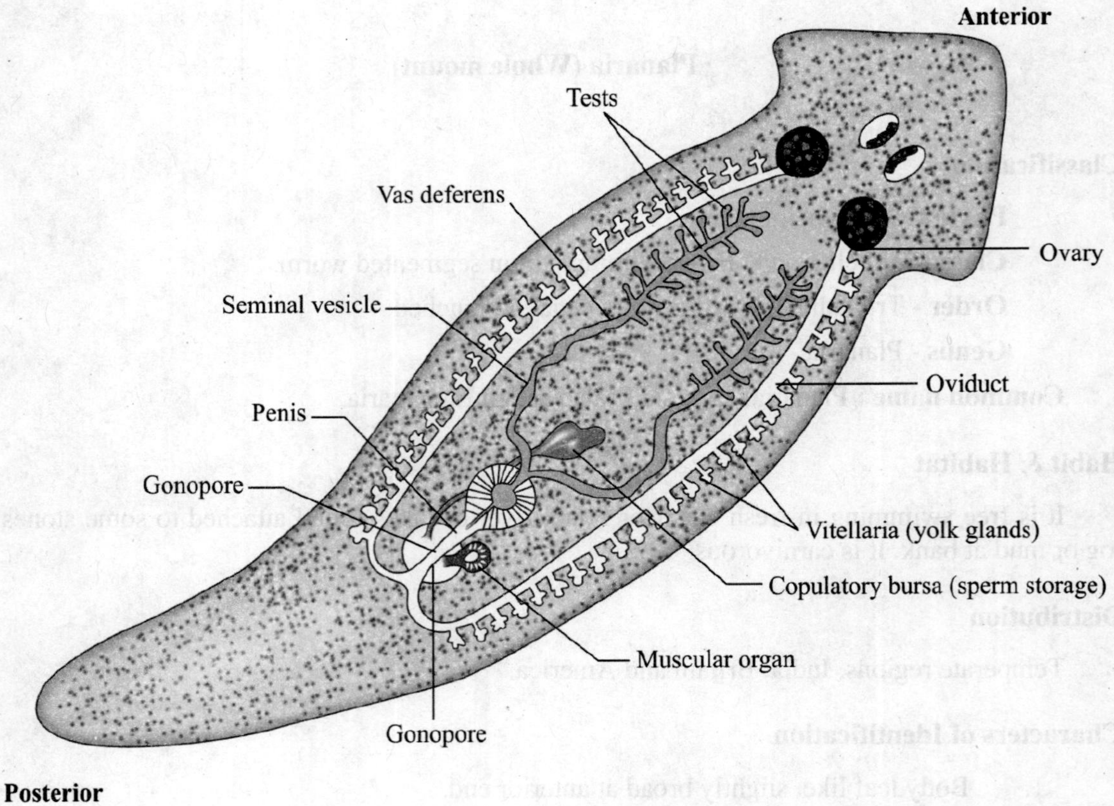

Fig. 1.39. Planaria

Fasciola (W.M.)

Classification

Phylum - Platyhelminthese (i) Flat worms

Class - Trematoda (i) Parasites

Order - Digenea (1) Two hosts in life cycle (2) Two suckers present

Genus- Fasciola

Common name: Known as *Fasciola hepatica* Liver fluke or sheep fluke worm.

Habit and Habitat

1. Lives as endoparasite in liver (bile ducts) on animals like sheep, horse, man, monkey, elephant etc.

Distribution - Cosmopolitan.

Characters of Identification

1. Leaf like flat body
2. Mouth surrounded by oval sucker & ventrally placed.
3. Between oral sucker (acetabulum) and mouth lies *gonopore.*

Special characters

1. Life cycle involvus two hosts i.e. *digenetic*
2. Snail, *Lymnea trancatula* is intermediate host and sheep as primary host.

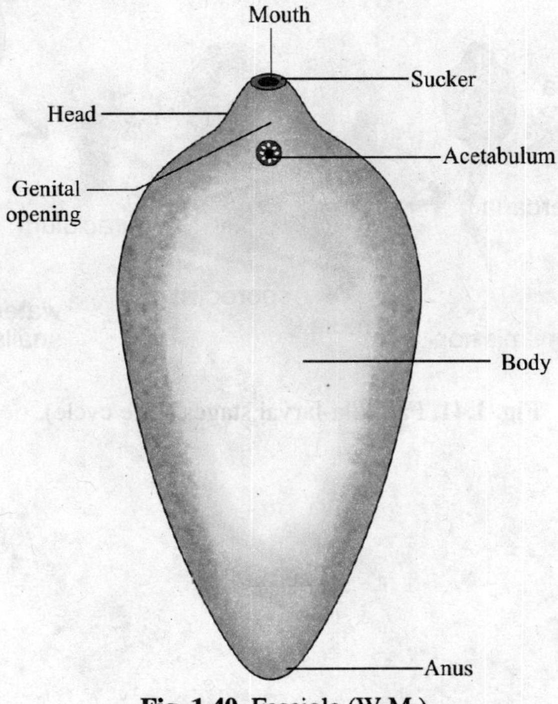

Fig. 1.40. Fasciola (W.M.)

Biological importance

1. Studied as Laboratory animal
2. It exhibit a biological phenomenon - *Polyembryony.* Five larval forms viz. *Miracidium, Sporocyst, Redia, Cercaria* and *Metacercaria* are found. Miracidium and Cercaria are free swimming stage.

Economic importance

1. It causes *liver rot* in sheep.
2. It causes Anaemia, Diarrhoea, Ulcers and Pain.
3. The sheep when infected reduce wool and meat.

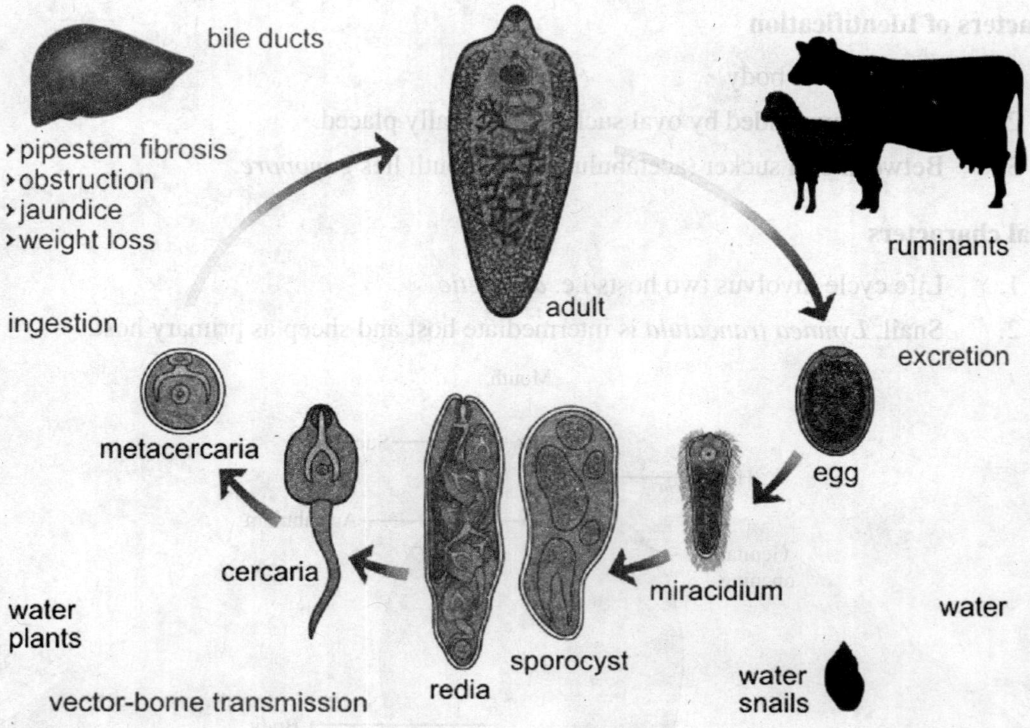

Fig. 1.41. Fasciola-larval stages (Life cycle).

Taenia

Classification

Phylum - Platyhelmithese

Class - Cestoda

(i) Intestine absent

Order - Cyclophillidea/Taenoidea

(i) Four suckers present on scolex.

Genus - Taenia (*Taenia solium*)

Common Name : Commonly called *Tape worm* or *Pork tape worm.*

Habit and Habitat: Endoparasite as in the intestine of man.

Distribution: Pork consuming countries.

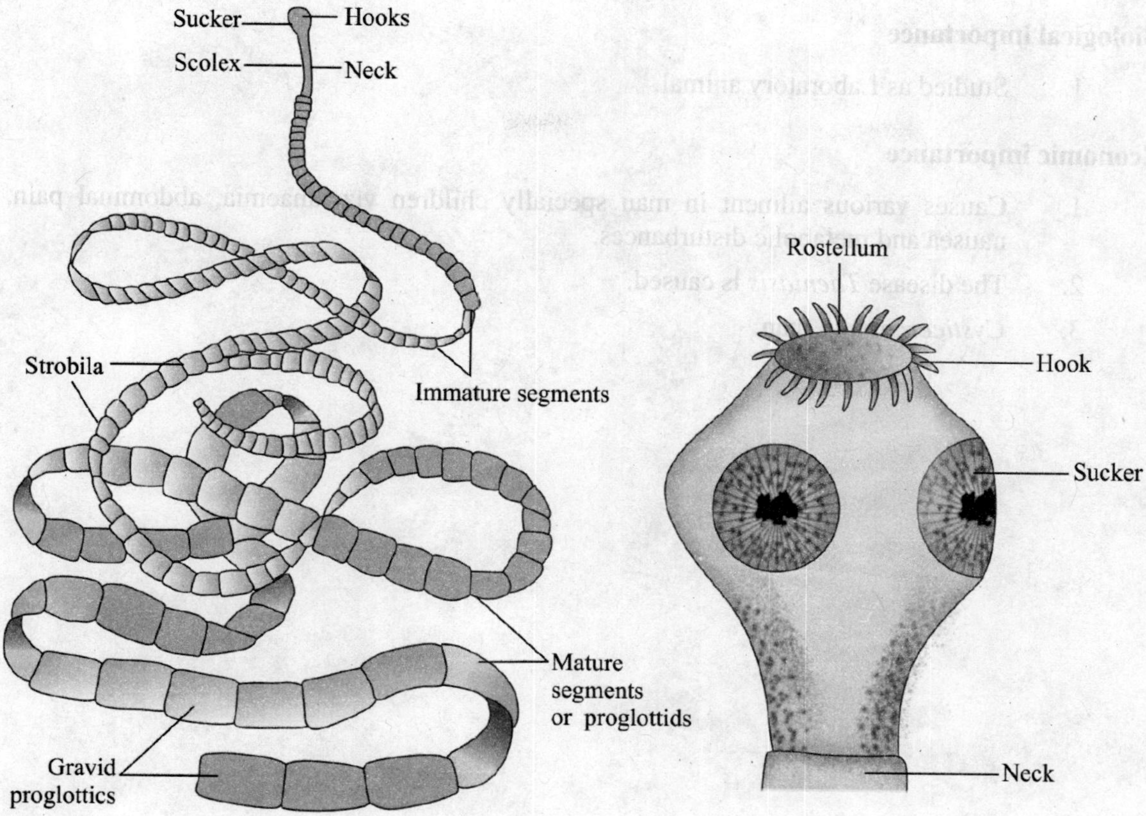

Fig. 1.42. Taenia, scolex (magnified)

Characters of Identification

1. Body consist of scolex, neck and segments.
2. Smaller segments are towards neck followed by mature and gravid segments. The segments are called *proglottids.*
3. Scolex contain four suckers and *rostellum,* 28-32 hooks are there in rostellum.

Special features

1. Mature proglottid is *hermaphroditic*
2. Gravid proglottid contains *hexacanth* that develops into *cysticercous* or *bladder worm.*
3. Life cycle is digenetic. Pig is *intermediate host* and man as *primary host*. Pig when consumes human excreta, receives *onchosphare* and become infected. They encyst in muscles of pig. It is called *measly pork.* Uncooked meat of pig when consumed by the man, the person become infected.
4. Enormous reproduction.

Biological importance

1. Studied as Laboratory animal.

Economic importance

1. Causes various ailment in man specially children viz. anaemia, abdominal pain, nausea and metabolic disturbances.
2. The disease *Taeniasis* is caused.
3. *Cysticercosis* in man.

Outline classification of Aschelminthese
(Askos - cavity, *helminths* worm)

Aschelminthese: Bilaterally symetrical unsegmented group of animals with pseudocoel and body covered with cuticle is known as *Aschelminthese.* It includes larger group *Nematoda* and smaller groups viz. *Rotifera, Gastrotricha, Kinorhyncha* and *Nematomorpha.* According to Hyman (1951), it is regarded as distinct phylum and these groups as class whereas some authors have considered these smaller groups as independent phylum.

Aschelminthese

Class

Nematoda | Nematomorpha | Rotifera | Gastrotricha | Kinorhyncha

Class- Nematoda

1. Aquatic, terrestrial or parasitic.
2. Cylindrical body tapering at both ends with cuticle.
3. Intestine is well formed.

Order - Ascaroidea

Example - Ascaris, Wuchereria, Trichinella

Order - Ascaroidea

1. Round worms. e.g. - Ascaris

Order - Trichuroidea

1. Whip worm. e.g. - Trichinella

Order - Filaroidea

1. Filarial worms
2. Pharynx muscular

Example - Wuchereria.

Order - Strongyloidea

1. Copulatory bursa present.

Example - Ancylostoma.

Order - Oxyuroidea

1. Pharynx with bulb

Example - Oxyuris, Enterobius

Order - Rhabditoidea

1. Parasitic or free living in soil

Example - Rhabditis.

Class - Namatomorpha

1. Thread like unsegmented body.
2. Adults free swimming, youngs are parasitic in crustaceans and insects.

Example - Gordius.

Class - Rotifera

1. Transparent, aquatic and microscopic.
2. Ciliary apparatus at anterior end.
3. Pharnyx with jaw apparatus called mastax
4. Flame cells present.

Example - Brachionus, Philodina

Class - Gastrotricha

1. Aquatic, fresh or marine water
2. Unsegmented worm like body with cilia in some parts.
3. Paired protonephridia present.

Example - Turbanella

Class - Kinorhyncha

1. Microscopic
2. Head retractile and with spines.

Example - Echinoderes.

Wuchereria

Phylum - Aschelminthese

Class - Nematoda

Order - Filaroidea

Genus - Wuchereria (*Wuchereria bancrofti*)

Common name : Commonly called *Filarial worm.*

Habit and Habitat : It is parasitic and adult worm found in lymphatic vessels.

Distribution : Tropical and temperate regions. In India in Bihar and Gorakhpur.

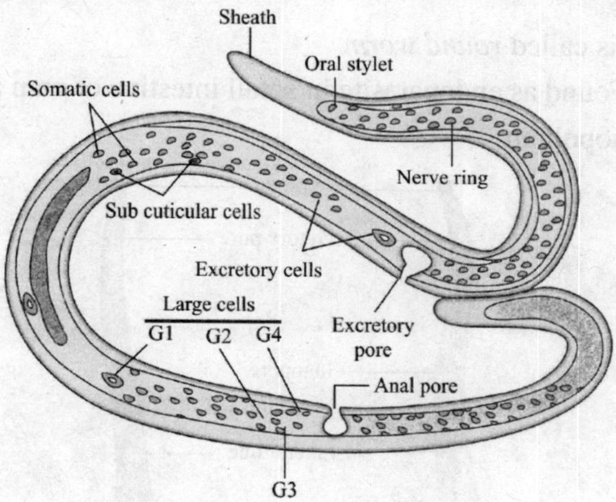

Fig. 1.43. Wuchereria

Characters of identification

1. The female worm give birth to *microfilariae* larva that resides into the blood vessels of man has eversible stylet at anterior end and tail is tapering.
2. It comes to the peripheral blood circulation during night. The female of *Culex fatigans* when suck the blood of man, receives them. It act as *vector.*
3. Microfilariae exhibits *diel* and nocturnal *periodicity*. Between 10 p.m. to 2 a.m. they remains in peripheral blood circulation. During day they remain in heart kidney and pulmonary vessels.
4. The lymphatic vessels are blocked and the legs become swollen. The disease is called *elephantasis* or *Filariasis.*

Biological importance

1. Studied as Laboratory animal.

Economic importance

1. It causes filariasis/Elephantasis
2. Causes Mental depression and headache also.

Ascaris

Classification

Phylum - Aschelminthese

(1) Pseudocoelomate

Class - Nematoda

(1) Commonly called round worms

Order - Ascaroidea

(1) Mouth with three lips

Genus - Ascaris

Common name - It is called *round worm.*

Habit & Habitat - Found as endoparasite in small intestine of man and other animals.

Distribution - Cosmopolitan

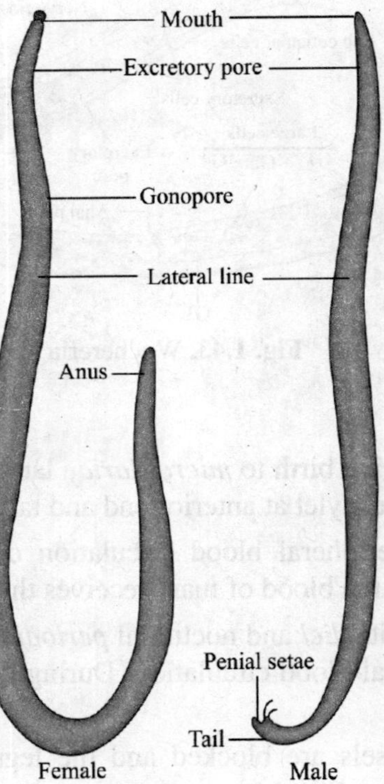

Fig. 1.44. Ascaris

Characters of identification

1. Creamy white or pink is colour, cylindrical body.
2. Male has coiled tail while female pointed.
3. Mouth is guarded by one dorsal and two sub-ventral lips.

External features or Sexual Dimorphism

	Male		Female
1	Body smaller, 15-30 cms. long and 3-5 mms. Wide.	1	Body larger, 20-40 cms, long and 6-8 mm, wide
2	Tail end curved ventral wards	2	Tail end straight.
3	Anus and genital pore common i.e., a cloacal aperture is present.	3	Anus and genital pore-separate. Genital pore midventral at about 1/3rd body length behind anterior end.
4	A pair of needle-like penial setae in cloaca	4	No penial setae
5	Pre-anal and post-anal papillae present to help in copulation	5	No such papillae

Special characters

1. *Sexual dimorphism* present. Male shorter (15-30 cm) than female (20-35 cm)
2. Presence of pseudocoelom
3. Segment (cuticle) present in skin.
4. Respiration by diffusion.

Biological importance

1. It is studied as a Laboratory animal.

Economic importance

It causes disease *Ascariasis,* abdominal pain, swelling and diarrhoea, children are commonly affected.

Dracunculus

Classification

Phylum - Aschelminthese

Class - Nematoda

Order - Dracunculoidea

Genus - Dracunculus (*Dracunculus medinensis)*

Common name : Commonly called *Guinae worm* or Naru

Habit and Habitat : The young one are found in freshwater and the adult in the skin of man.

Distribution: Africa, India, Egypt and UAE.

Characters of identification and special character

1. The female is about 1 meter large and male about 4 cm only.
2. The mature female erupts through a blister (of toxic secretions of worm) on those parts of the body that generally comes in contact with water e.g. legs, arms. The uterus releases micro larvae on contact with water. These micro larvae are eaten by cyclops in the fresh water.
3. When a person drinks unfiltered water, the cyclops reaches to man and infect it.

Economic importance

1. It causes *Dracunculiasis*

The female comes out from a painful blister slowly 2. The infected person feel acute pain and can vomit sometimes. The efficiency of man is affected and discontinued from his work for few days that cause severe economic losses to him.

Special comment

The infection is common in the areas where the people used stepwell's water for drinking. Pratapgarh, Udaipur, Banswara, Dungarpur and Jhadol regions (tribal belt) of Rajasthan were mainly affected. The efforts of UNICEF has eradicated this disease in Rajasthan. Stepwells were closed and special boltine silk cloth was distributed to filter the water.

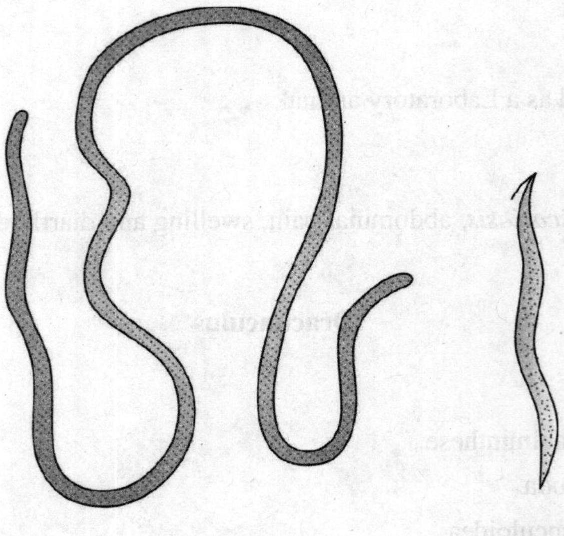

Fig. 1.45. Dracunculus

Outline classification of Annelida

Annelida (Annulus-ring eidos-worm)

- Class
  - Archiannelida
    e.g. Polygordius
  - Polychaeta
    - Order
      - Errantia
        e.g. Neries
      - Sedantaria
        e.g. Arenicola chaetopterus
  - Oligochaeta
    - Order
      - Limnocolae
        e.g. Tubifix chaectogaster
      - Terricolae
        e.g. Pheritima Lumbricus
  - Hirudinea
    - Acanthobdellida
      e.g. Acanthobdella
    - Rhynchobdellida
      e.g. Branchalion
    - Gnathobdellida
      e.g. Hirundinaria

Neries

Classification

Phylum : Annelida

(1) Segmented cylindrical body

(2) Chitinous setae present

Class : Polychaeta

(1) Many setae

(2) Parapodia-well developed

Order : Errantia

(1) Pharynx protrusible

(2) Jaw chitinous

Genus : Neries

Common name : Commonly called *Sand worm, Clam worm* or *Rag worm.*

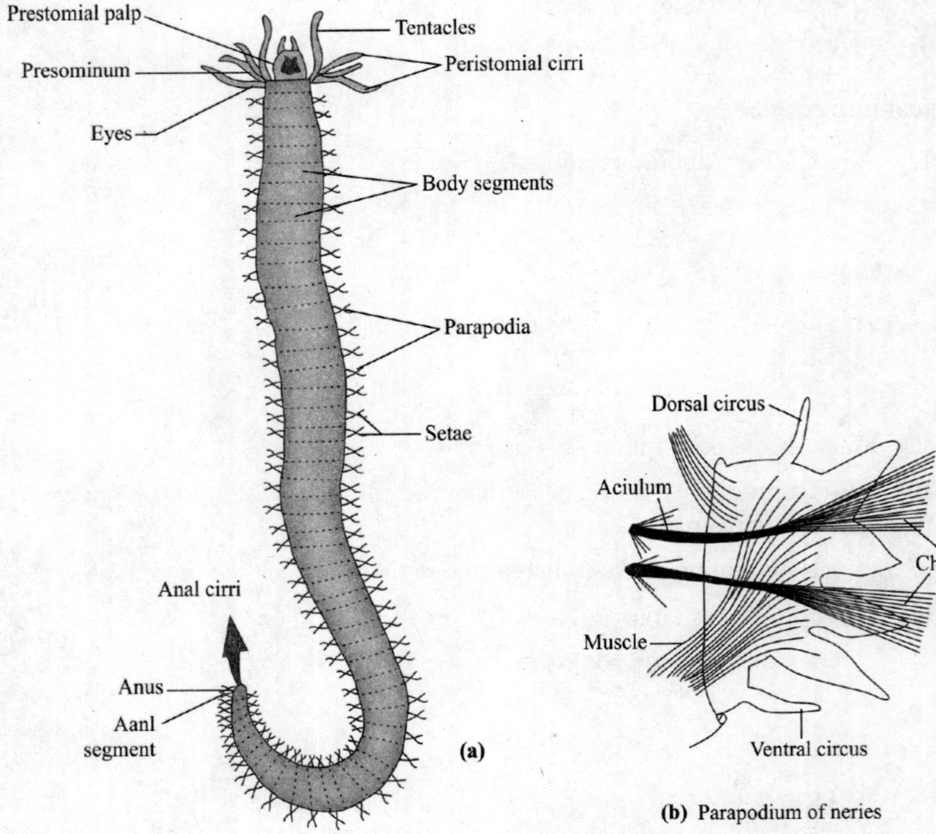

Fig. 1.46. a. Neries, b. Parapodium of Neries

Habit and Habitat

1. Burrowing in sea bottom, free swimming also
2. Nocturnal and carnivorous

Distribution: Cosmopolitan

Characters of identification

1. Body about 8 inch long with 80-120 segments.
2. Anterior part divided into *Prostomium* and *Periostomium.*
3. Prostomium bears 2 pair of eyes, one pair of tentacles and palp. Periostomium bears a pair of cirri.
4. A pair of parapodia on each segment present.

Special characters

1. It is called *Heteroneries* in reproductive phase (also called *epitok*)
2. Cleavage spiral
3. *Trochophore* larva is present.
4. *Autotomy* and *regeneration* is found.

Biological importance

1. Studied as Laboratory animal.
2. It is a component in Marine food chain.

Economic importance

It is used as bait for anglar fishes.

Heteroneries

(Classification and description as Neries).

1. It is reproductive phase of Neries. Body is divided into *atoke* and *epitoke*. Epitoke is reproductive part.
2. Free swimming phase (whereas the neries crawls at the bottom of sea)
3. Eyes becomes large
4. Parapodia are large and their setae becomes oar shaped.

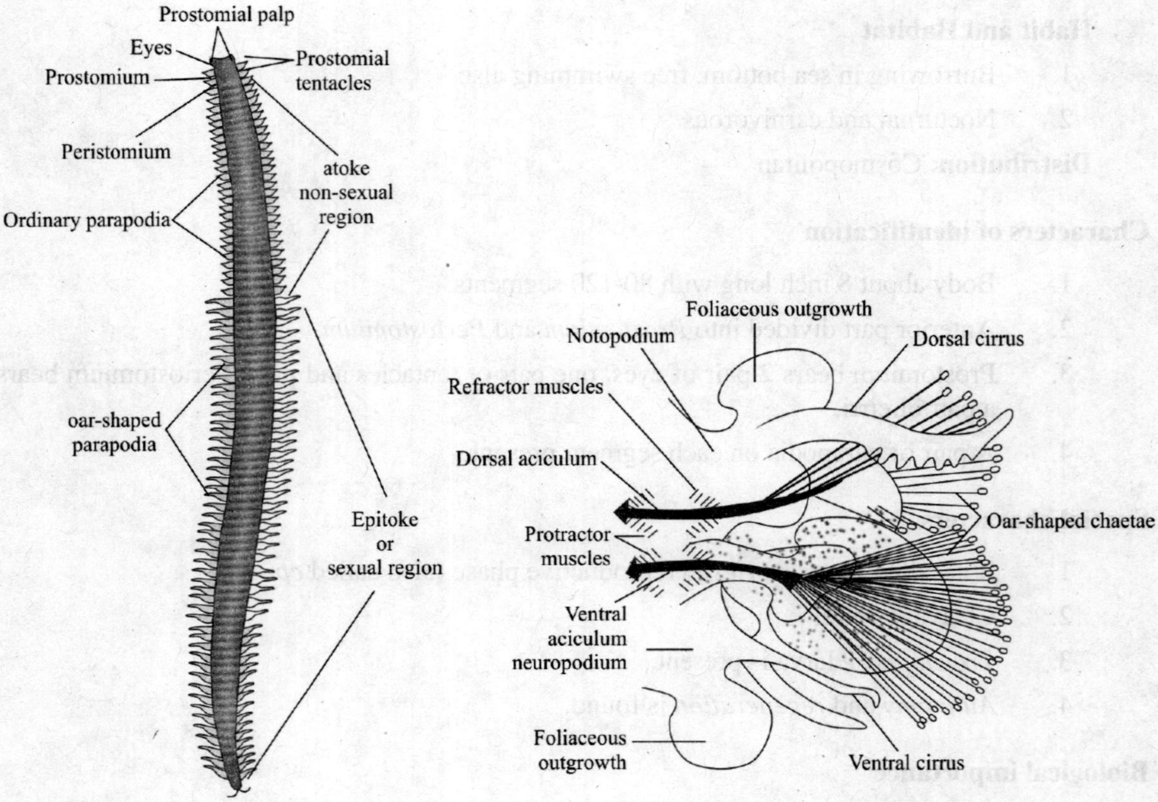

Heteroneries (a) **Parapodia of heteroneries (b)**

Fig. 1.47. Heteroneries

Unit - 2
Anatomy

1. EARTHWORM (*Metaphire posthuma*)

Earthworm is classified in class *oligochaeta* under *Phylum Annelida* which is known for the *metamerism* (external and internal segmentation). They possess *true coelom.*

External Features

1. Body cylindrical made up of more than 100 segments.
2. Lives in moist and shaded places.
3. Nocturnal in habit.
4. Mouth located at the anterior tip and anus on posterior. The *Prostomium* covers the mouth.
5. Clitellum from XIV to XVI (14-16) segments. It is saddle shaped swollen glandular area responsible for the production of cocoon.
6. Mid dorsal line extends from one end to the other.

Note : Earlier name *Pheretima posthuma* has been changed. It is *Metaphire posthuma* now. *Metaphire* (= *Pheretima*) *posthuma*.

7. Female reproductive pore one, in the middle of the 14th segment, the male reproductive opening two, on the lateral side of the 18th segment.
8. Numerous minute nephridiopores present all over the body except segment I and II.
9. Each segment (except first and last) bears bristle like setae located ventrolaterally which aids in locomotion.
10. Earthworm is *hermaphroditic* or Bisexual.

Internal Anatomy

1. Take the preserved earthworm and lay it in dissecting tray exposing its dorsal surface. Locate the mid dorsal line.

2. Slit open by making a small cross section just posterior to clitellum and then insert the tip of the scissors in anterior direction upto prostomium.

3. Fold back the flaps of the body wall and pin them. The pin should be slanting. Study the internal anatomy as per instructions of your teacher. Locate the Nerve ring, Alimentary canal, Seminal vesicles, Dorsal & Ventral blood vessels, Nerve cord etc.

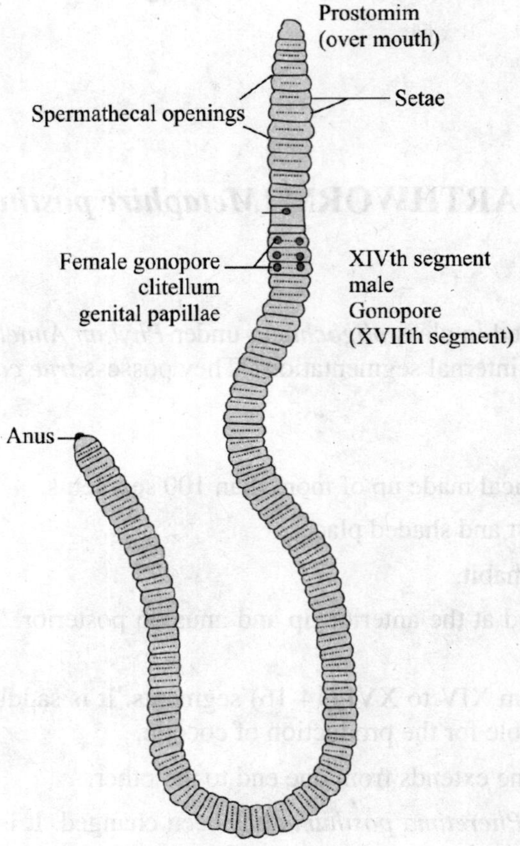

Fig. 2.1 : Earthworm

NERVOUS SYSTEM AND NERVE RING

1. Open the earthworm as described above.
2. Displace the alimentary canal in the pre-clitellar area.
3. Nerve cord is visible below the intestine. Trace the nerve cord upto IVth segment. Nerve ring is visible as yellow band.
4. Cut the pharynx above to nerve ring and loose the muscles. Push the pharynx below from the ring and locate the nerve ring showing *sub-pharyngeal ganglia, circumpharyngeal connective* and *supra-pharyngeal ganglia.*

   The dorsal part of the ring is *supra-pharyngeal ganglia*. The lateral part is *circum pharyngeal connective* and ventral part is sub *pharyngeal ganglia.*

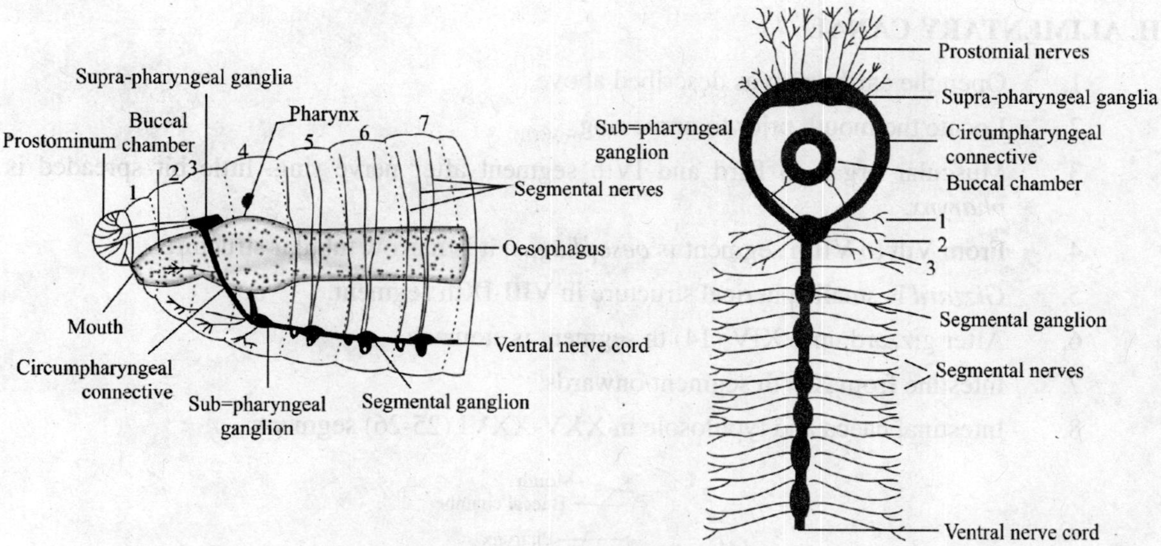

Fig. 2.2 (a) Nervous system (b) Nerve ring

5. Locate and trace the nerve cord upto last segment. There is a ganglia is each segment (from Vth segments onword) and 3 parts of segmental nerves from each ganglia are visible.

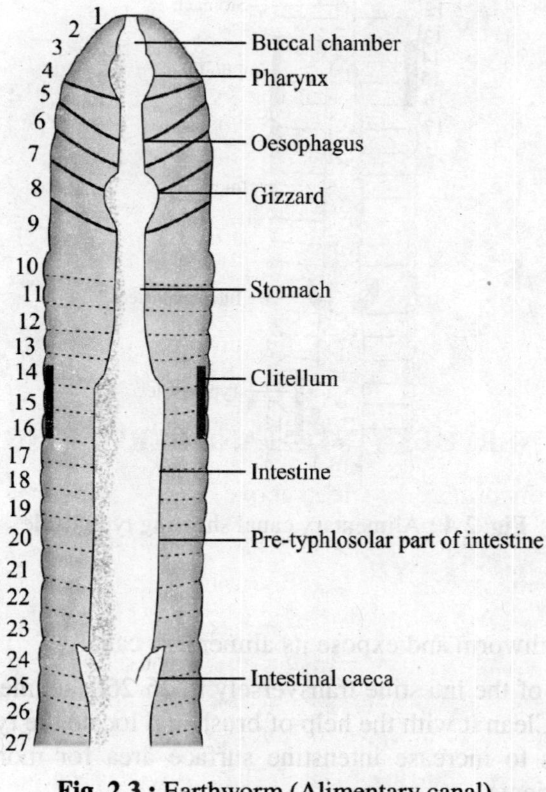

Fig. 2.3 : Earthworm (Alimentary canal)

II. ALIMENTARY CANAL

1. Open the earthworm as described above.
2. Locate the mouth prior to nerve ring.
3. Muscular organ in IIIrd and IVth segment after nerve ring, little bit spreaded is *pharynx.*
4. From Vth to VIIth segment is *oesophagus*, it is narrow tubular structure.
5. *Gizzard* is small spherical structure in VIII-IXth segment.
6. After gizzard upto XIV (14) th segment is stomach.
7. Intestine from XIVth segment onwards.
8. Intestinal caeca and typhlosole in XXV-XXVI (25-26) segments.

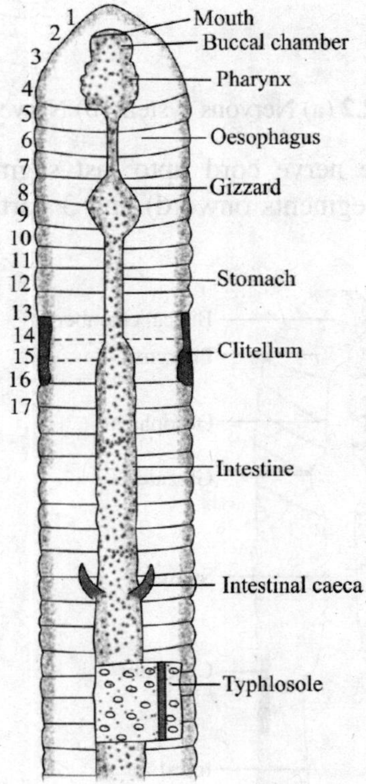

Fig. 2.4 : Alimentary canal showing typhlosole

Typhlosole

1. Open the earthworm and expose its alimentary canal.
2. Cut the wall of the intestine transversely in 25-26th segment and unfold the cut part in one side. Clean it with the help of brush and locate the typhlosole. The function of typhlosole is to increase intenstine surface area for more efficient absorption of digested nutrients.

REPRODUCTIVE SYSTEM (Male and Female)

1. Open the earthworm as described above.
2. Locate the *prostate glands* in post clitellar part, *Seminal vesicles* are located XI-XIIth (11-12) segment and spermathecae in VI-IXth (6-9) segments.
3. Testis are two pairs in X-XI th (10-11) segments and spermathecal funnel also two pairs in the same segment. *Vas deferense* are parallel to nerve cord upto prostate glands.
4. Ovaries are two pairs located at the junction of XII-XII th (12-13) segments. Oviducal funnel in XIIIth segment below each ovary.

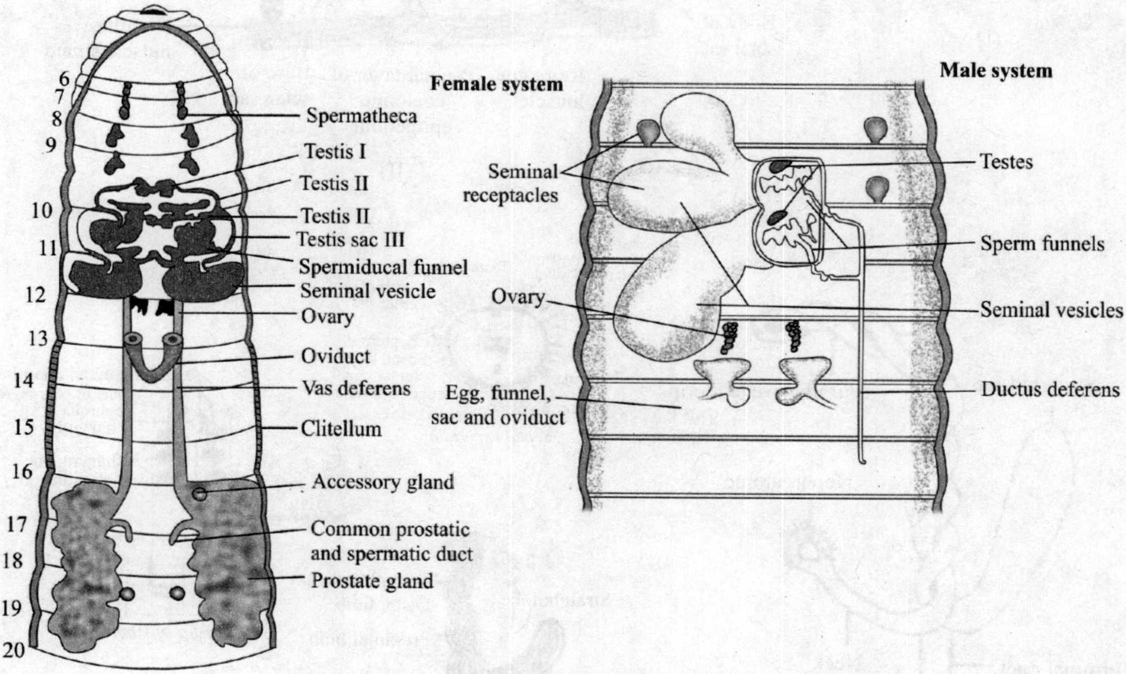

Fig. 2.5 : Earthworm : Reproductive system (in situ)

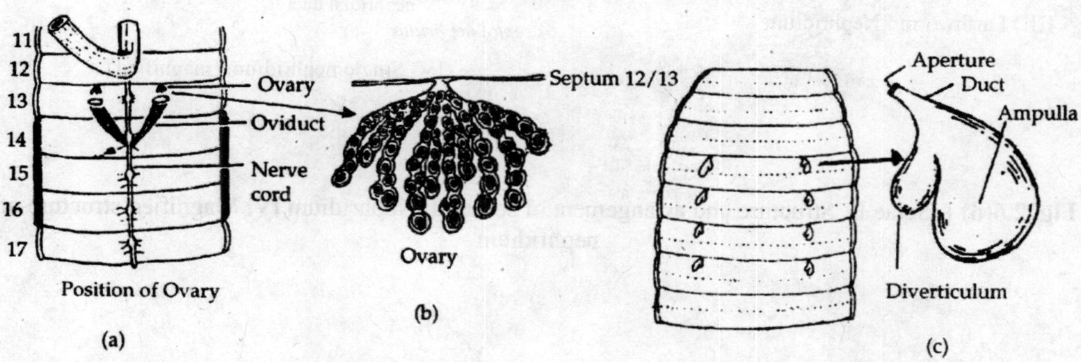

Fig. 2.6(a) Earthworm : Reproductive system simplified presentation (b) Ovary (c) Spermatheca

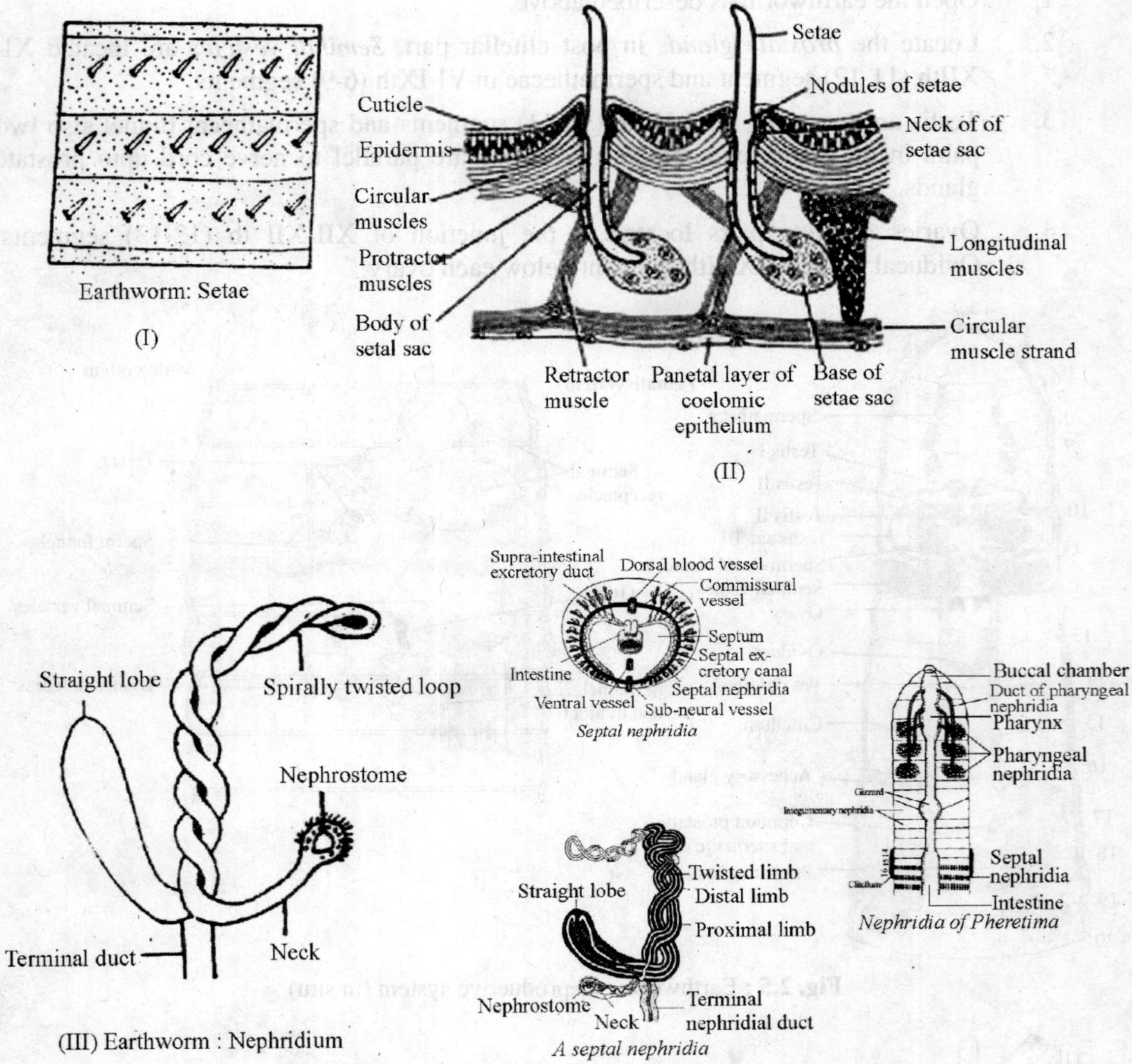

Fig. 2.6(d) I. Setae II. Structure and arrangement of setae III. Nephridium IV. Magnified structure of nephridium

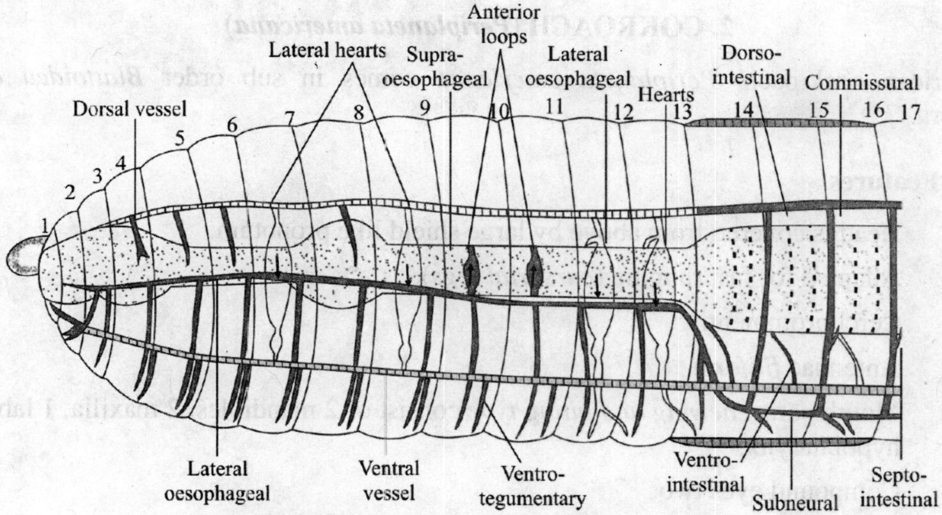

Fig. 2.7 : Earthworm : Circulatory-system

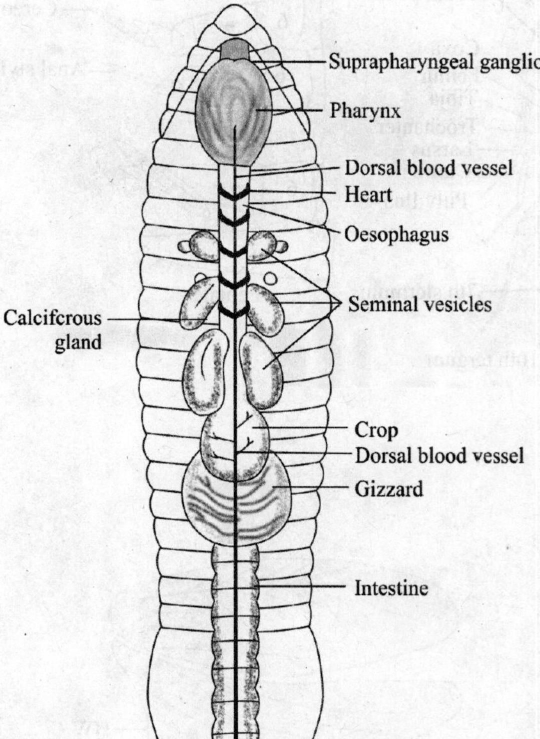

Fig. 2.8 : Semi-diagrammatic showing internal anatomy of *Lumbricus*

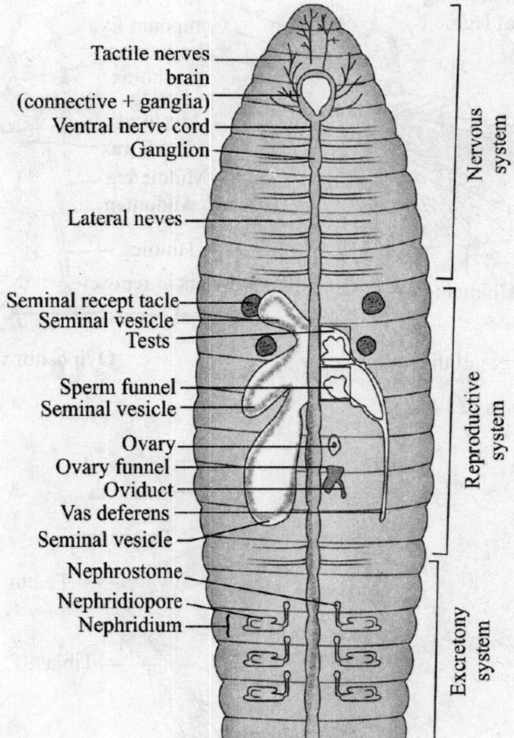

Fig. 2.9 : Semi-diagrammatic showing nervous, reproductive and excretory system in *Lumbricus*

2. COKROACH (*Periplaneta americana*)

American cockroach, *Periplaneta americana* comes in sub order *Blattoidea of* order Dictyoptera.

External Features

1. Head is covered from above by large shield like pronotum.
2. Adapted for fast running (*cursorial legs*)
3. Cerci prominent.
4. Antennae *filiform*
5. Mouth parts *chewing and biting type* consist of 2 mandibles, 2 maxilla, 1 labium & 1 hypopharynx.
6. Compound eyes two.
7. Nocturnal in habit and *omnivorous.*

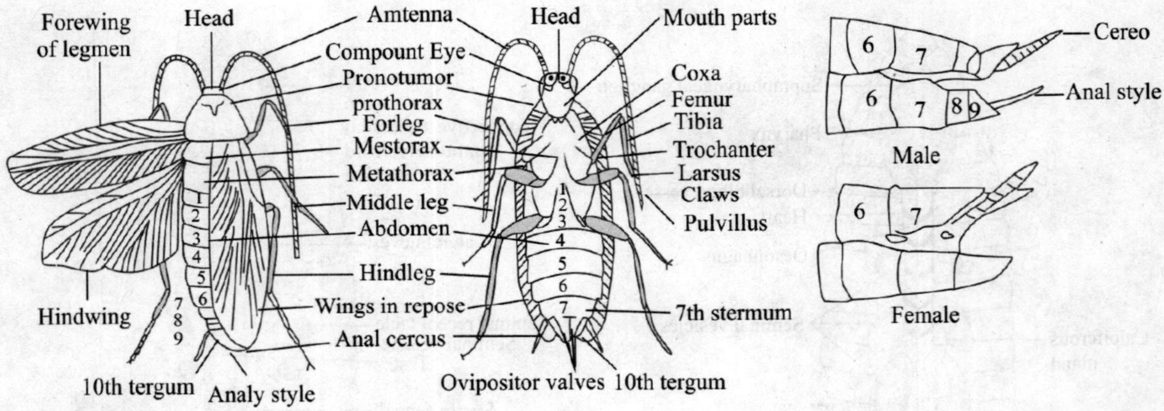

Wings & LEG

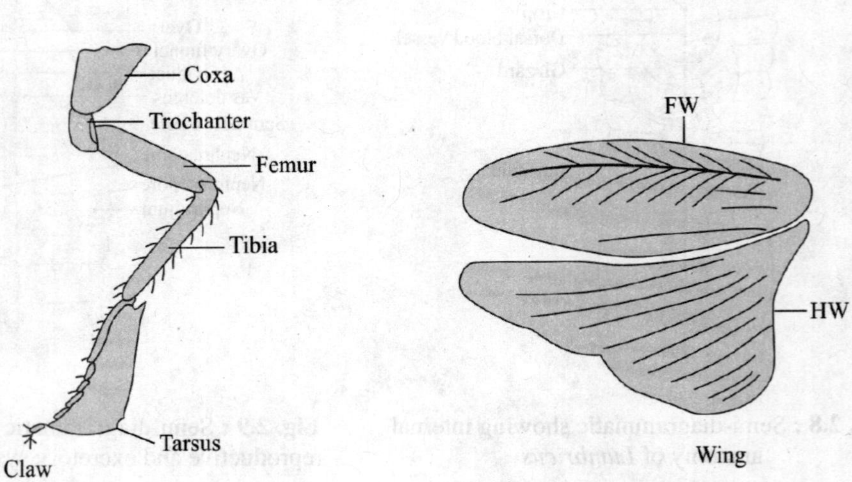

Fig. 2.10 : Cockroach : (a) external features : Dorsal and ventral side, (b) Wings and leg

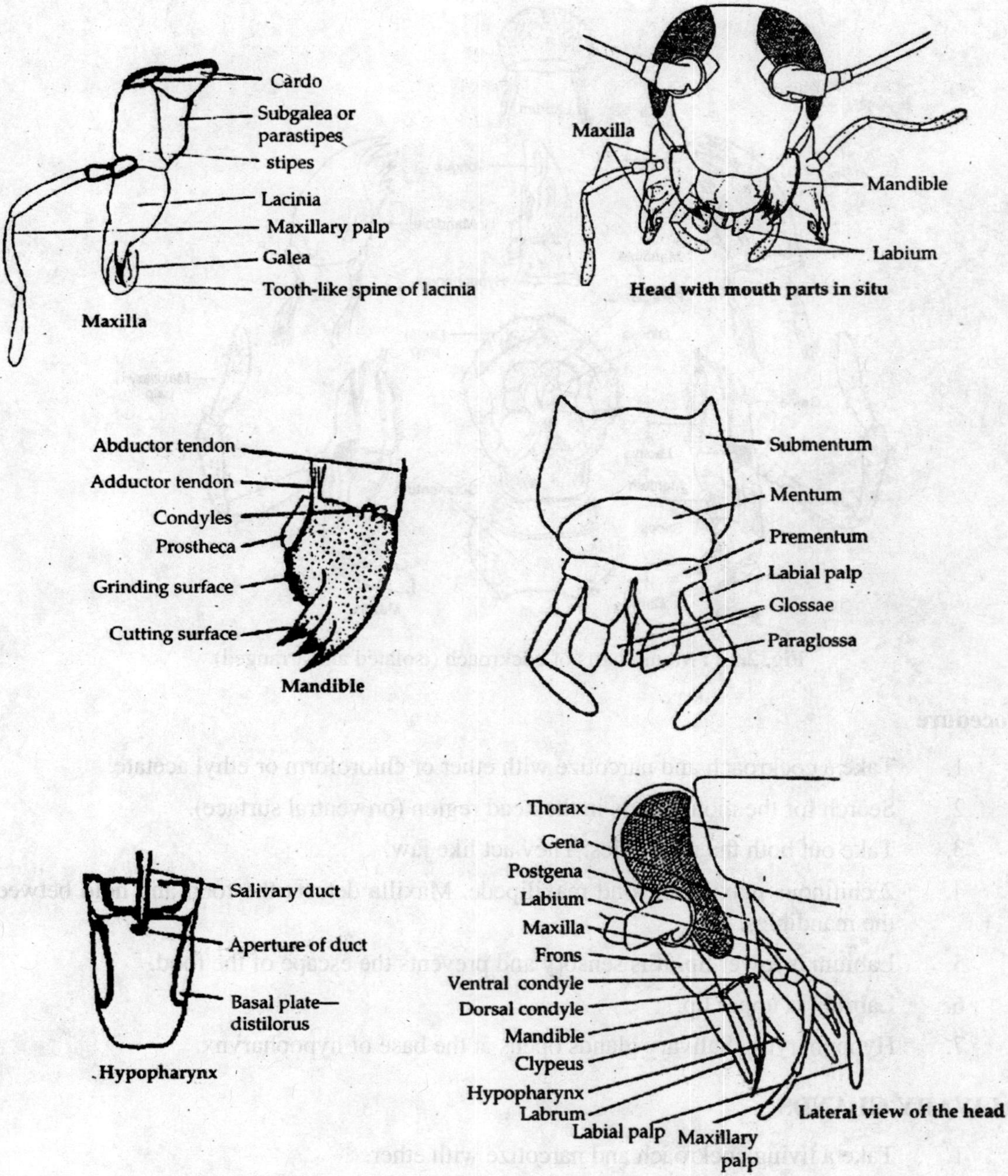

Fig. 2.11 Head and mouth parts of cockroach

MOUTH PARTS

They include mandibles, maxilla, labium, labrum and hypopharynx.

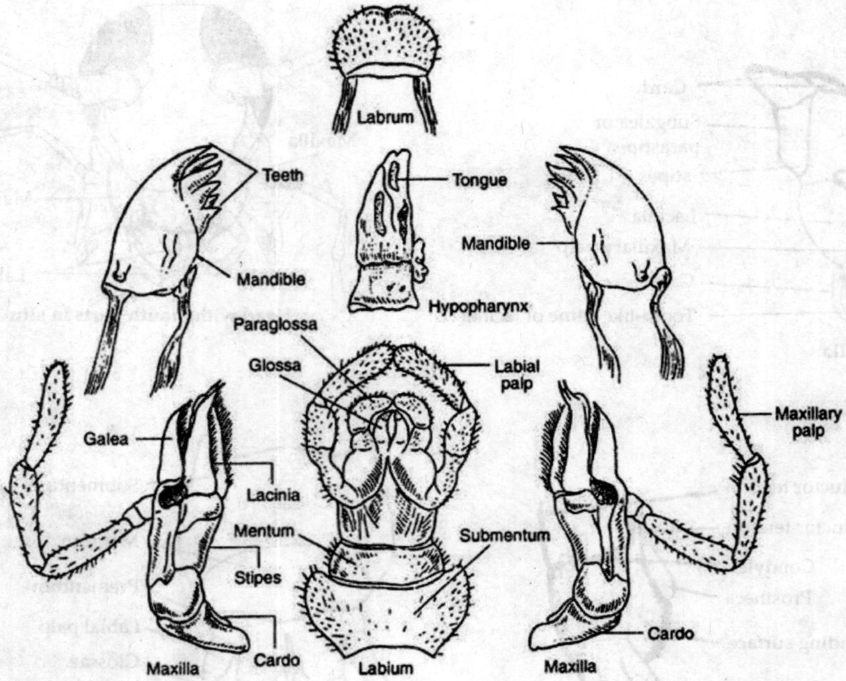

Fig. 2.12 : Mouth parts of cockroach (isolated and arranged)

Procedure

1. Take a cockroach and narcotize with ether or chloroform or ethyl acetate.
2. Search for the mouth parts in the head region (on ventral surface).
3. Take out both the mandibles. They act like jaw.
4. 2 chitinous maxilla beyond maxilipede. Maxilla detects the food and hold between the mandibles.
5. Labium is lower lip. It is sensory and prevents the escape of the food.
6. Labrum is upper lip.
7. Hypopharynx, Salivary glands opens at the base of hypopharynx.

SALIVARY GLANDS

1. Take a living cockroach and narcotize with ether.
2. Cut the pleura of both side.
3. Fix the cockroach in the tray with the help of pins in its head and femur.
4. Now remove the tergal plates.
5. Expose the salivary glands and take out into the watch glass.
6. It consist of reservoirs and glands. Both of them opens through ducts at the base of hypopharynx.

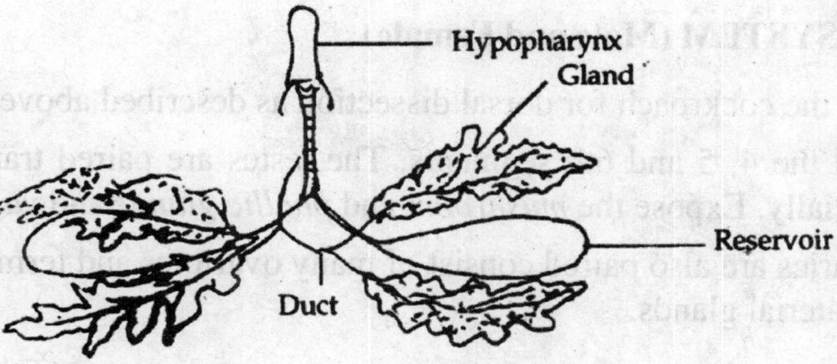

Fig. 2.13 : Salivary Glands (Isolated)

ALIMENTARY CANAL

1. Procure living cockroach and narcotize with ether vapours. Fix in the dissecting tray exposing its dorsal surface.
2. Cut the pleura of both sides.
3. Remove the tergal plates.
4. Expose the alimentary canal. It consist of salivary glands, oesophagus, crop, gizzard, midgut, hepatic caeca, Malpighian tubules, ileum, colon and rectum.

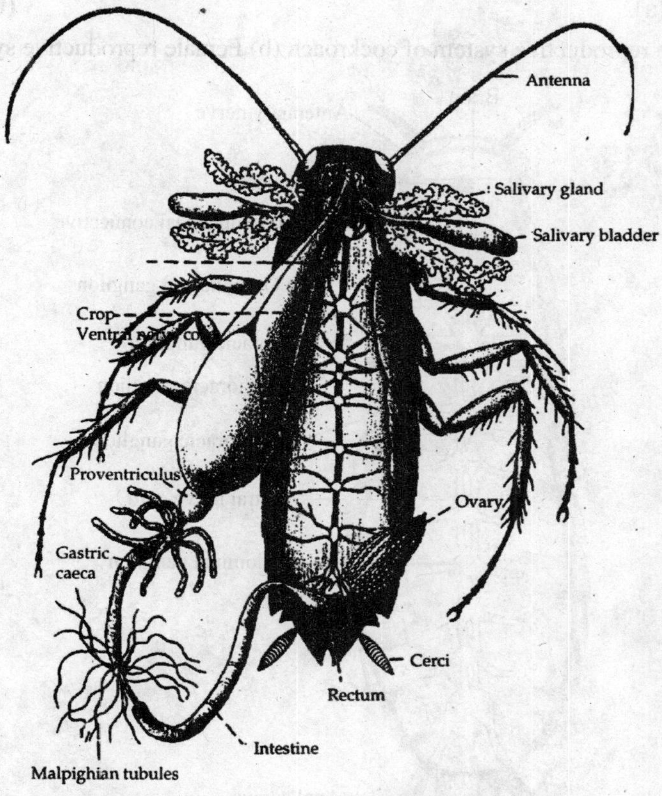

Fig. 2.14 : Cockroach : Alimentary Canal (in situ)

REPRODUCTIVE SYSTEM (Male and Female)

1. Prepare the cockroach for dorsal dissection as described above.
2. Look in the 4, 5 and 6th segments. The testes are paired transparent lobes located superficially. Expose the *mushroom* and *phallic glands* and *vas deferens.*
3. The ovaries are also paired consist of many ovarioles and terminal filaments. Expose the collaterial glands.

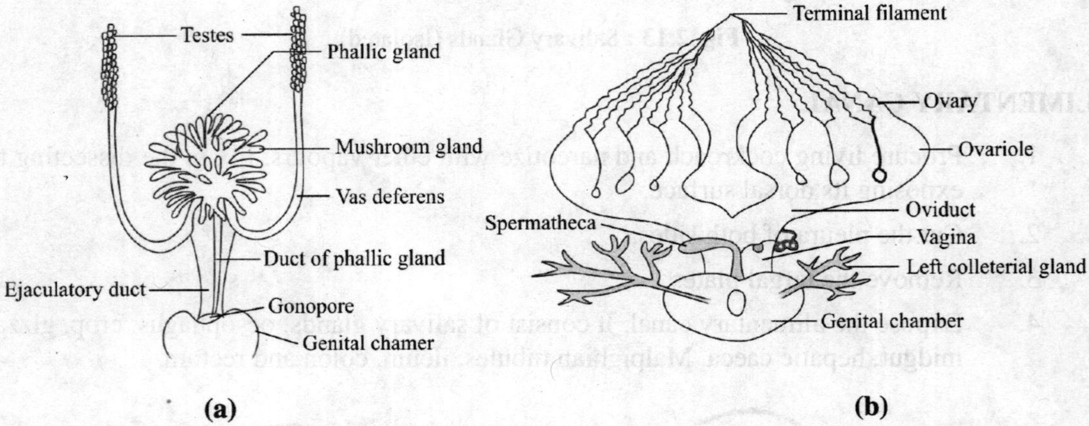

Fig. 2.15 (a) Male reproductive system of cockroach (b) Female reproductive system of cockroach

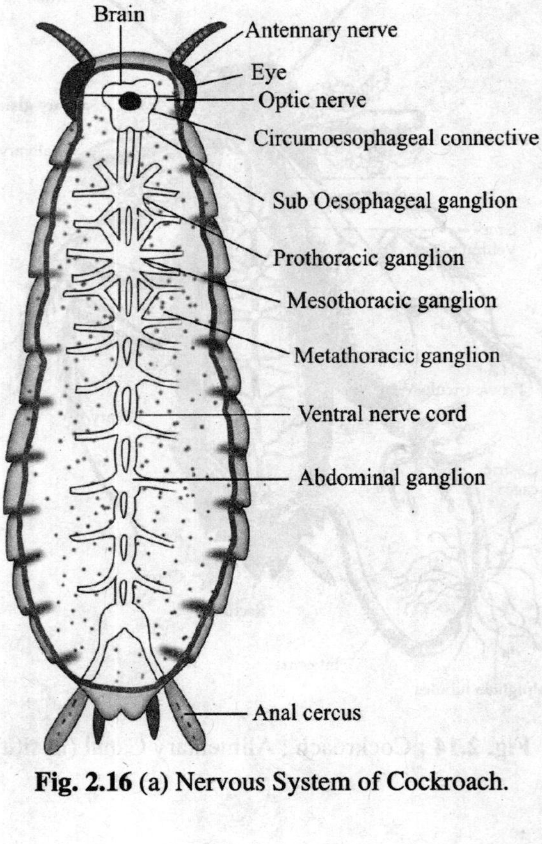

Fig. 2.16 (a) Nervous System of Cockroach.

Nervous System

1. Prepare the cockroach for dorsal dissection as described above.
2. Remove the alimentary canal and locate the nerve cord.
3. Trace the ganglia in each segments upto 6th segment and its nerves.
4. Trace the Thoracic ganglia also.
5. Now remove the, head sclerites carefully and expose the brain. It is a bilobed structure.

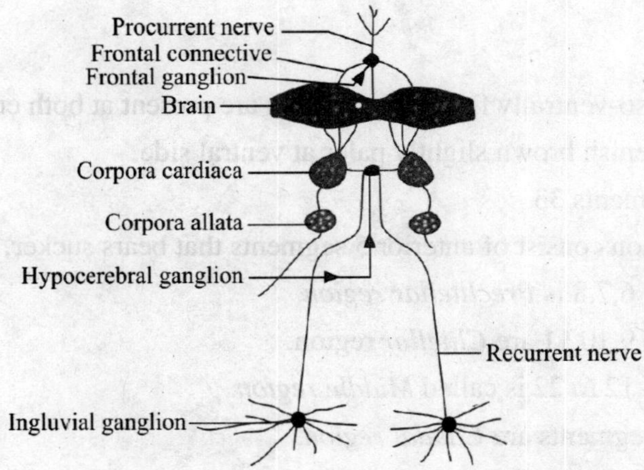

Fig. 2.16 (b) Stomodeal sympathetic nervous system (Cockroach)

Endocrine complex (diagrammatic)

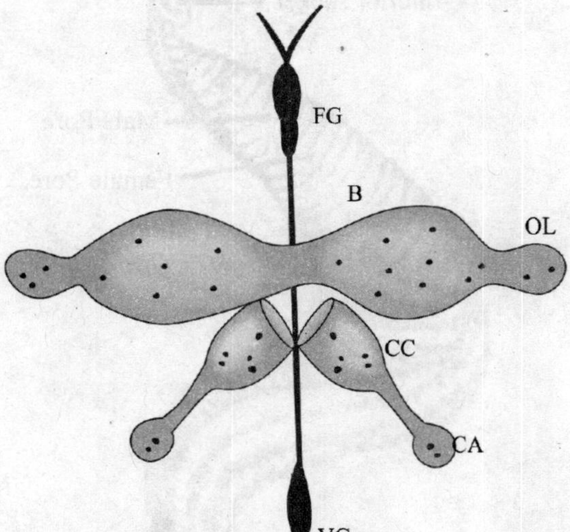

Fig. 2.17. Endocrine complex of cockroach. FG - Frontal ganglia, B - Brain, OL - Optic lobe, CC - Corpora cardiaca, CA : Carpora allata, VG - Visceral ganglia

Mounting of Salivary Gland

1. Prepare the cockroach for dorsal dissection and expose its salivary glands.
2. Take out the glands in a watch glass and dehydrate in alcohol series and mount in DPX.
3. For temporary mounting - mount in glycerine.

3. HIRUDINARIA (LEECH) INDIAN CATTLE LEECH

Salient features

1. Body dorso-ventrally flattened. Suckers are present at both ends.
2. Body greenish brown slightly paler at ventral side.
3. Body segments 33
4. Head region consist of anterior 5 segments that bears sucker, mouth, jaw and eyes.
5. Segments 6,7,8 is *Preclitellar region.*
6. Segments 9,10,11 are *Clitellar* region.
7. Segments 12 to 22 is called *Middle region.*
8. 23 to 26 segments are *Caudal region.*
9. Segment 26 to 33 is *Posterior region.*
10. Male genital opening in 10th segment and female 11th segment.

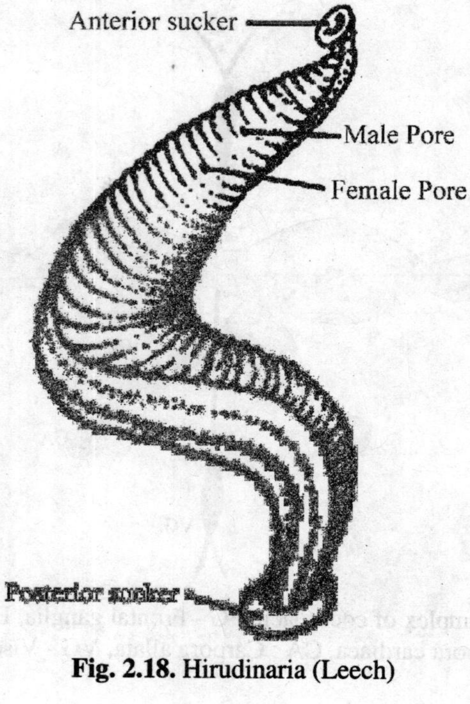

Fig. 2.18. Hirudinaria (Leech)

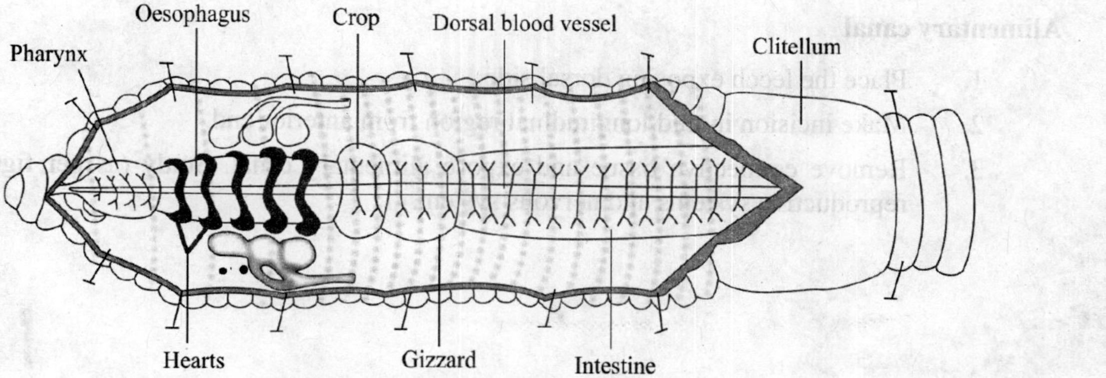

Fig. 2.19. Leech: Alimentary canal

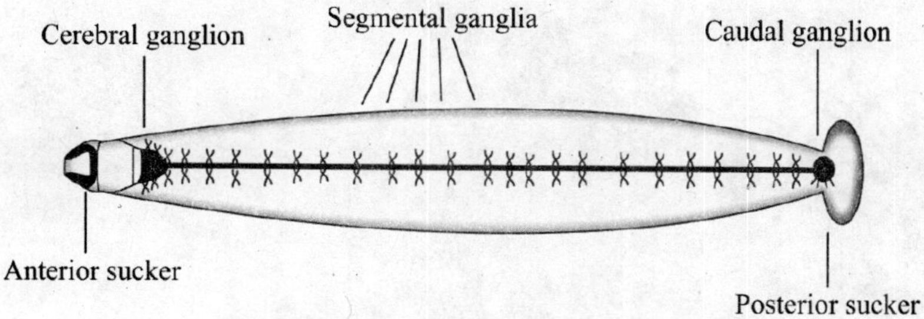

Fig. 2.20. Leech: Nervous system

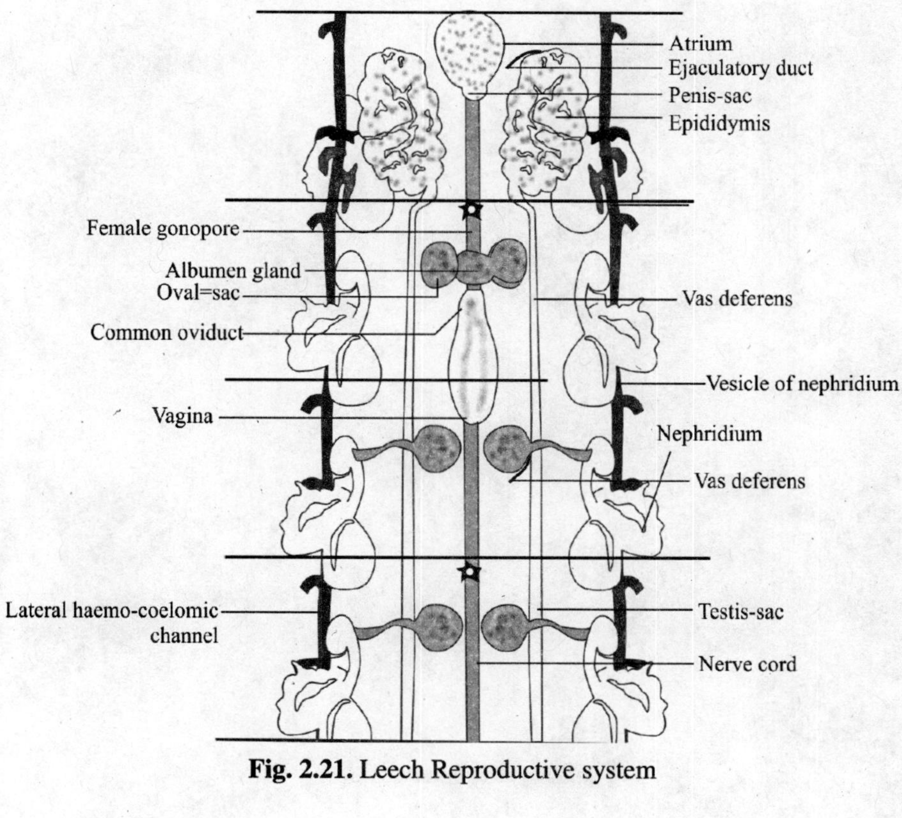

Fig. 2.21. Leech Reproductive system

Alimentary canal

1. Place the leech exposing dorsal side.
2. Make incision in mid longitudinal region from anterior end.
3. Remove connective tissue and expose alimentary canal. Study (as per figure) the reproductive systems and nervous system.

■■■■

Unit - 3
Permanent Preparations

1. PERMANENT SLIDES

1. Paramecium (binary fission)
2. Paramecium (conjugation)
3. Foraminifera shell
4. Sycon (L.S.)
5. Sycon (T.S.)
6. Spicules
7. Spongin fibres
8. Gemmule
9. Medusae
10. Daphnia (W.M.)
11. Cyclops (W.M.)
12. Miracidium
13. Sporocyst
14. Redia
15. Cercaria
16. Metacercaria
17. T.S. Body of Fasciola
18. T.S. mature Proglottid and Gravid proglottid
19. Leech, testicular nephridia

Paramecium (Binary fission)

1. Binary fission is a mode of asexual reproduction.
2. Micro-nucleus divides mitotically.
3. Paramecium divides by a transverse constriction into two daughter paramecia.
4. Each daughter paramecium receives equal part of the nuclear apparatus and other contents.

Paramecium (Conjugation)

1. Conjugation is a mode of sexual reproduction.
2. It involves the union of two mating types each called conjugant, for the exchange of nuclear material. Conjugants attach themselves by their oral surfaces.

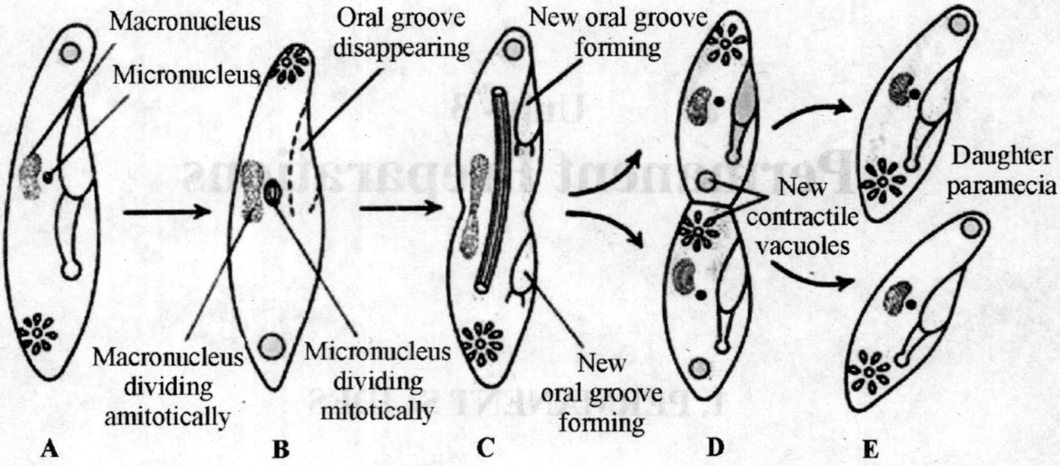

Fig. 3.1 Binary fission

3. Macronuclei breaks off into pieces & later absorbed by cytoplasm, micronuclei divides mitotically into 4 haploid nuclei. Three of them disappears. Left nuclei divides mitotically into two unequal gammete nuclei.
4. Small active is called *migratory gamete* nucleus and large inactive *gamete nucleus*.
5. Migratory gamete nucleus of each conjugant migrate to fuse with the stationary gamete nucleus to form a *zygote* or *synkaryon.*

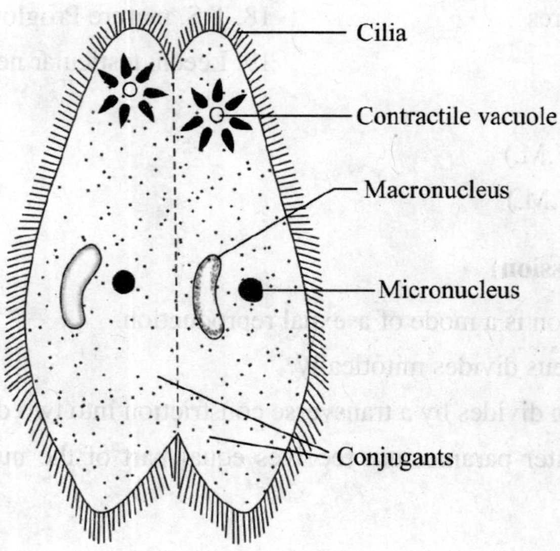

Fig. 3.2 Conjugation

Spicules

1. Spicules make up the skeleton of sponges.
2. *Calcareous* (made up of $CaCO_3$), *Siliceous* (made up of silica).

3. Large spicules are megasclers and small are microscleres.
4. On the basis of their axis and rays they are *Monaxons, Triaxons, Tetraxons, Polyaxons* and *Spheres.*

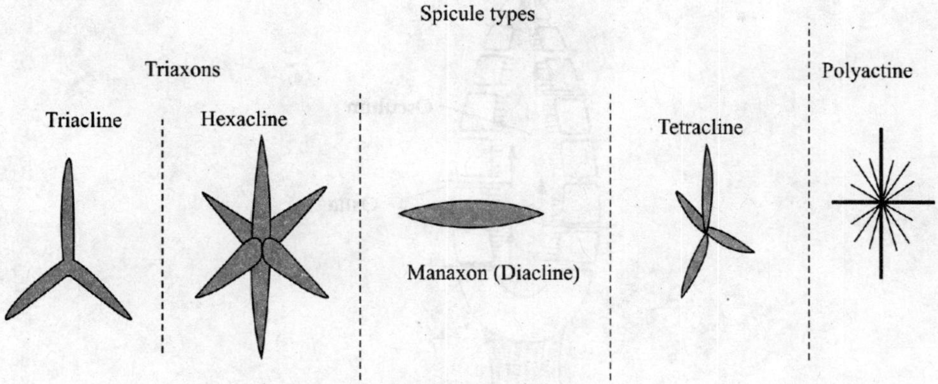

Fig. 3.3 Spicules

Spongin Fibres

1. Skeleton of horny sponges (*Keratosa*) e.g. *Euspongia* is made up of spongin fibres.
2. *Spongin* is secreted by mesenenchyme cells called *spongioblasts.*
3. They appear as irregular branched fibres.

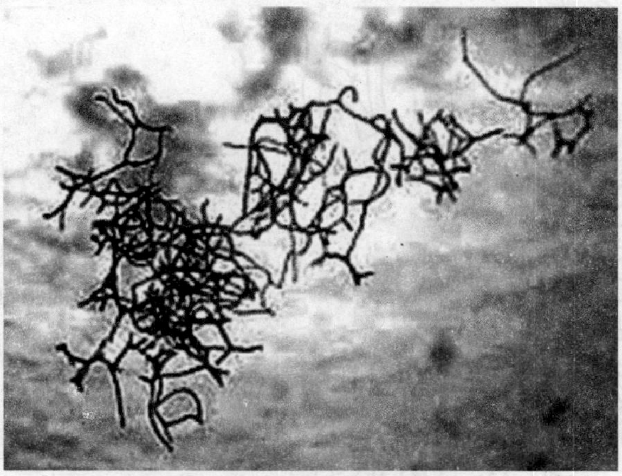

Fig. 3.4 Spongin fibres

Sycon (L.S.)

1. L.S. of Sycon exhibits incurrent canal which opens through ostia.
2. Incurrent canal is lined by *pinacocytes.*
3. Flagellated chamber is lined by *Choanocytes.*
4. Calcareous spicules are embedded in the body wall and tissues.
5. Sycon exhibits syconoid type of canal system.

Ostia $\xrightarrow{\text{incurrent canal}}$ Prosopyles $\longrightarrow$ Radial canal $\rightarrow$ Internal ostia $\rightarrow$ Spongocoel $\xrightarrow{\text{Osculum}}$ Outside

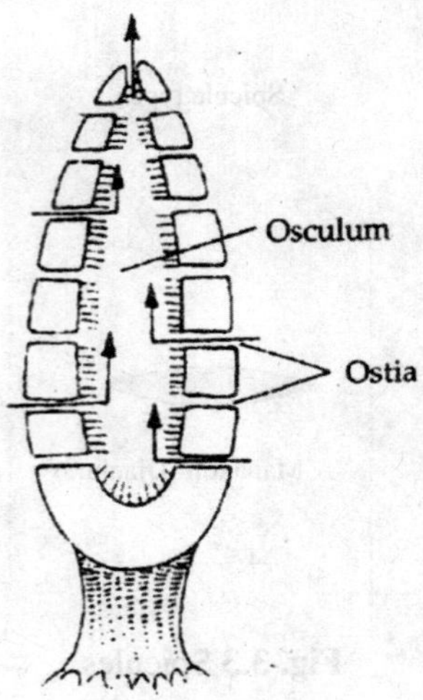

Fig. 3.5 Sycon L.S.

Sycon T.S.

1. Circular in section shows radial canal, ostia, incurrent canal, prosopyle spongocoel.
2. Spongocoel is lined by *pinacocytes* and flagellated chamber by *choanocytes*.

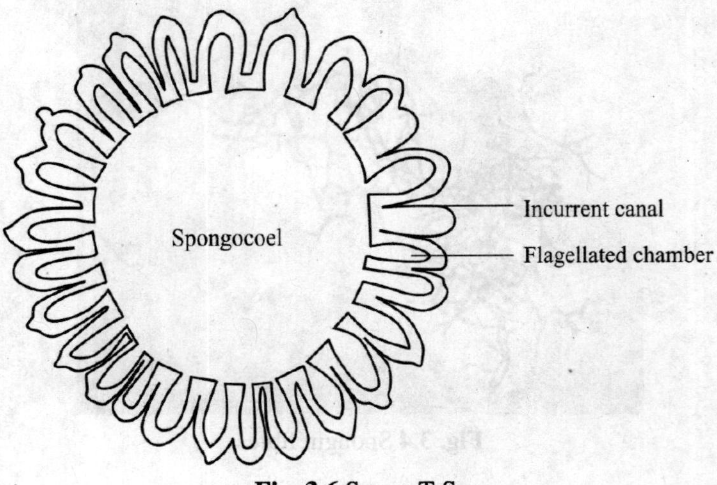

Fig. 3.6 Sycon T.S.

Foraminfera Shell

1. The dead members of order *Foraminifera* whose shell is made up of $CaCO_3$ collected at the bottom of the sea, called *foraminiferan ooze.*
2. The empty shell of polystomella and globigerina and other foraminifera are found deposited in the sea bottom. Huge bottom deposits of the chalk are obtained.

Fig. 3.7 Foraminiferan shells

Gemmule

1. Fresh water sponges (and some marine) during unfavourable conditions (cold and hot) form internal buds called *gemmule.*
2. Food laden (glyco-protein + lipoprotein) amoebocytes are enclosed in a two layered capsule.
3. The covering is made up of monaxon spicules and amphidiscs.
4. Micropyle present.

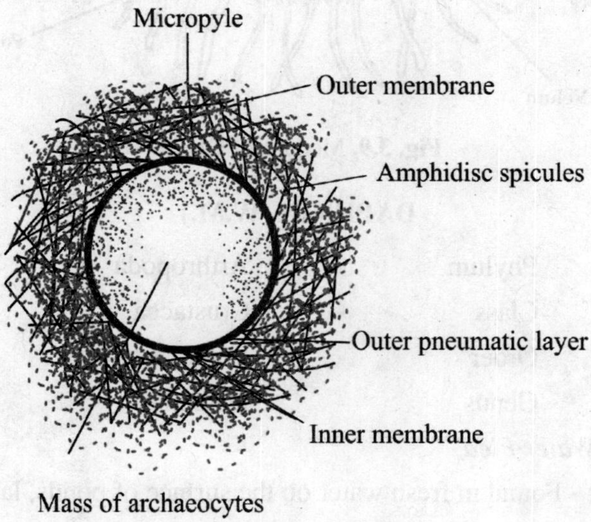

Fig. 3.8 : Gemmule

Medusae

1. It is *saucer* shaped.

2. It is solitary free swimming sexual zooid of obelia.
3. Outer convex surface is called ex umbrellar and inner concave called sub-umbrellar.
4. Manubrium in the centre of sub umbrellar surface present.
5. Tentacles at the rim of umbrella present.
6. 4 radial canals present.
7. 4 gonads present.
8. Velum present.

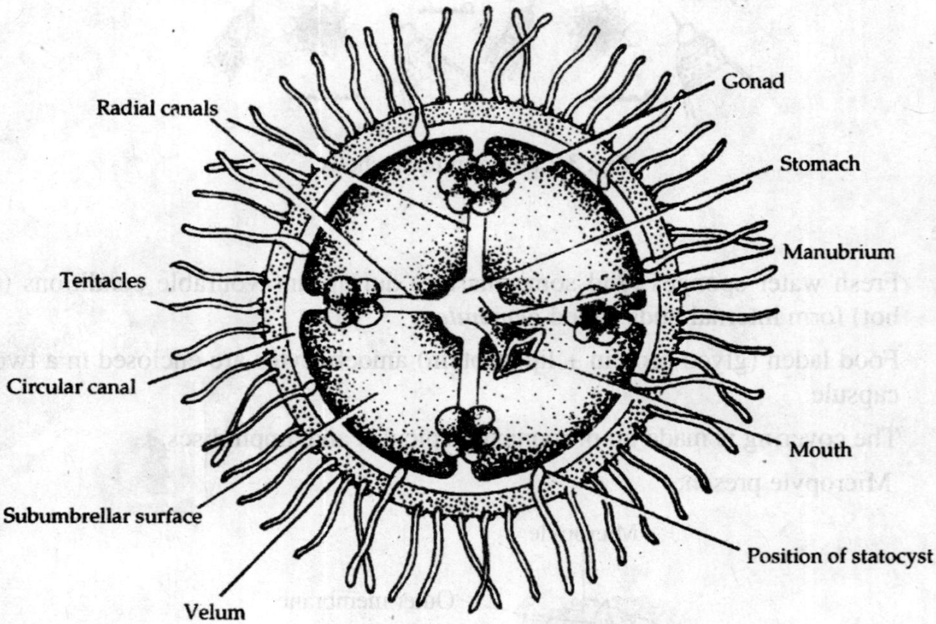

Fig. 3.9. Medusa of Obelia

DAPHNIA (W.M.)

Classification	Phylum	-	Arthropoda
	Class	-	Crustacea
	Order	-	Cladocera
	Genus	-	Daphnia

Common Name - *Water Flea*

Habit and Habitat - Found in fresh water on the surface of ponds, lakes and dams.

Characters of Identification

1. Soft laterally compressed body enclosed in a bivalved carapace which terminate into caudal spine.

2. Body divided into head and trunk.
3. Head with short beak, a pair of compound eyes, a pair of biramus antennae.
4. Thorax with 5 pair of appendages.
5. Female has a large brood pouch.

Special Features

1. They are *zooplanktons* and have important role in aquatic food chain.
2. It exhibit *Cylomorphosis* (different size in different seasons).
3. Shows *diel periodicity.*

Biological Importance

1. It is studied in the Laboratory as representative cladocera.
2. They are used as food in pisciculture.

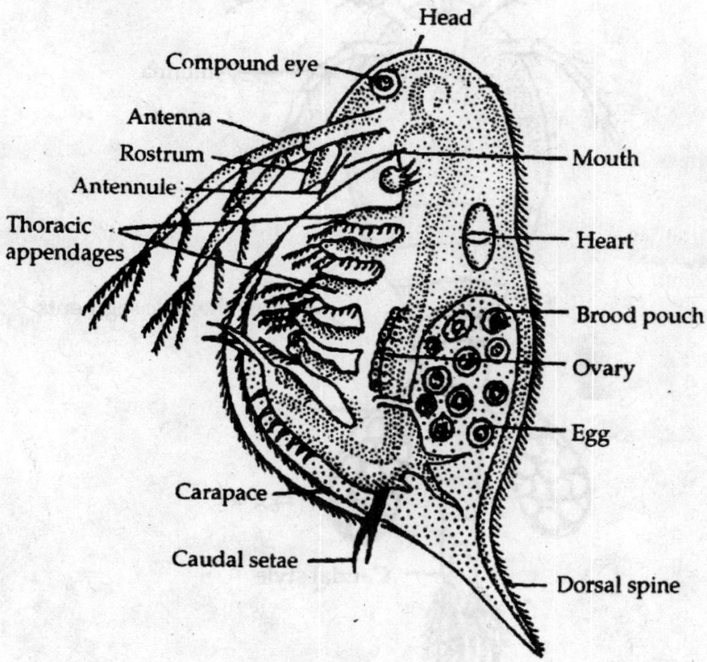

Fig. 3.10: Daphnia

CYCLOPS (W.M.)

Classification	Phylum	-	Arthropoda
	Class	-	Crustacea
	Order	-	Cyclopoida
	Genus	-	Cyclops

Common Name - Cyclops (also water flea)

Habit & Habitat - Found in fresh water pond, lakes, dams and ditches.

Characters of Identification

1. Body divided into cephalothorax and abdomen.
2. Head and thoracic segments are fused to form a cephalothorax. Head bears a single median eye, a pair of uniramous antennule and a pair of antennae.
3. Cephalothorax is followed by thoracic segments.
4. Telson which terminate into Caudal styles present.
5. Female carry a pair of egg sac.

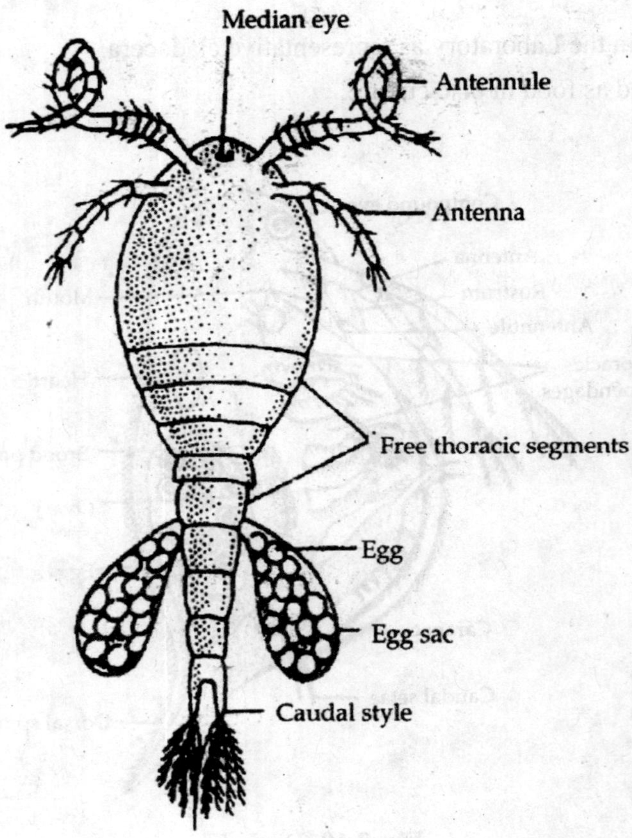

Fig. 3.11. Cyclops

Special Features

1. It acts as a vector for *Dracunculus medinensis* (guinea worm).
2. The juvenile *Dracunculus* is eaten by cyclops. When the person drinks unfiltered water, cyclops reaches to the intestine and release the Dracunculus.

Biological Importance

1. It is studied in the Laboratory as representative crustacea.

Economic Importance

It helps in the spread of disease *Dracunculiasis* commonly called *Naru* or *Bala.*

Additional Information

UNICEF has launched an extensive programme to eradicate Naru. It has been eradicated from the Rajasthan now. Earlier Dungarpur, Pratapgarh and Banswara districts were badly affected.

T.S. Body of Fasciola

1. Cuticle is outermost layer that bears backwardly directed spines.
2. Below cuticle is basement membrane.
3. Basement membrane lined by circular muscles to its outside and longitudinal muscles in side.
4. Coelom absent mesenchyma cells are filled.
5. Caeca, excretory canal, ovary, oviduct, uterus, shell glands and vitelline glands are visible in the section.

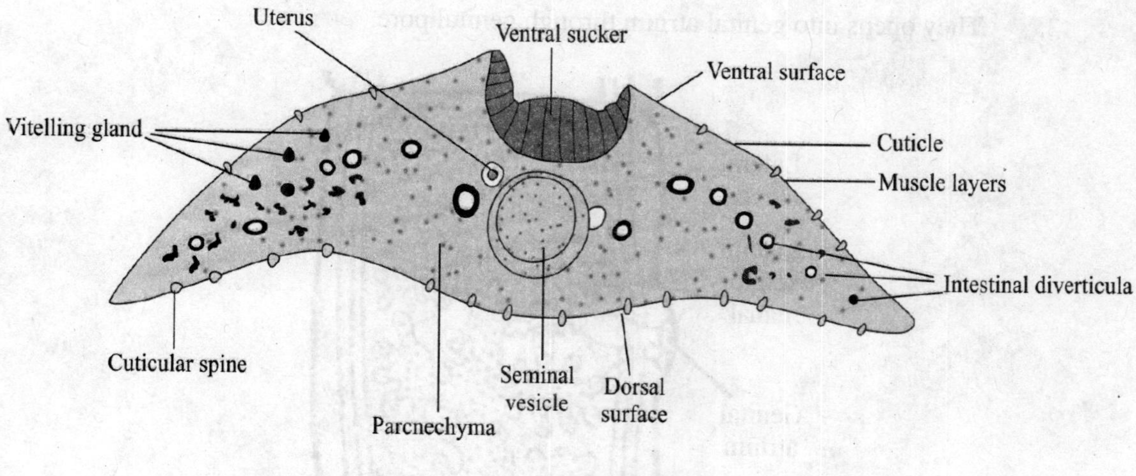

Fig. 3.12.

T.S. Mature proglottid of Taenia

1. Body is covered by cuticle. Below it are located longitudinal and circular muscles.
2. Coelom absent. Parenchyma tissue is filled in the space.

3. Nerve cord, excretory canal, ovary, oviduct, uterus, testis are visible in the section. Eggs are filled in most of the part.

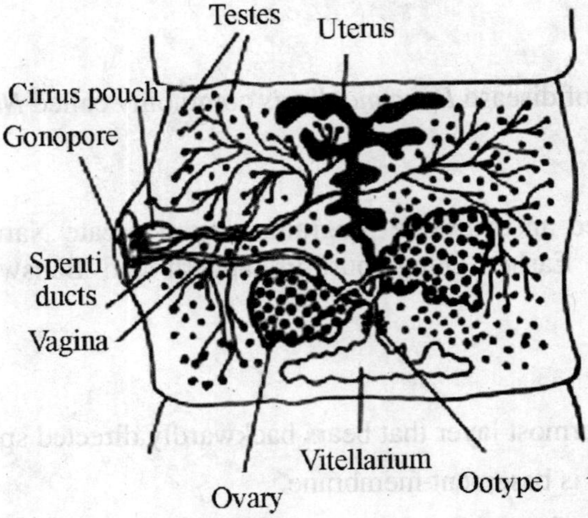

Fig. 3.13. Mature proglottid

Gravid Proglottid

1. Both male and female reproductive organs are found.
2. The male reproductive system consist of testis Vas deferens and cirrus and female a pair of ovaries, oviduct, vagina, uterus, shell gland and citellar membrane.
3. They opens into genital atrium through genital pore.

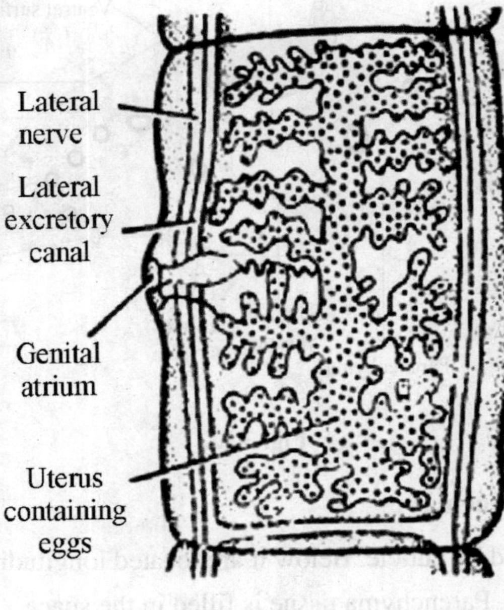

Fig. 3.14. Gravid proglottid

Cysticercus

1. It is second stage larva of Taenia. First is *hexacanth.* 3 pair of hooks are found (hence hexa, means 6). It is covered by membrane called *onchosphere*. Onchosphere only develops further if the excreta of man is taken up by the pig otherwise it dies.
2. When pig consumes excreta of man, it contains mature proglottids also. The proglottids are digested releasing *hexacanth.* Hexacanth reaches to striated muscles through blood circulation. Here it develops into *Cysticercus* or *bladder worm* (size - 10 mm). It is infective stage.
3. Its shape is like a bladder. An invagination at anterior end that develops rostellum, hook and sucker to its inside edge. Later its evagination takes place resulting these structures on scolex appear to its outer surface. It further does not develops until it reaches to man (primary host).
4. When a person eats uncooked pork. It reaches to stomach where bladder is digested and scolex stick to the intestine (with the help of hook and suckers). There, it grow into mature tapeworm.

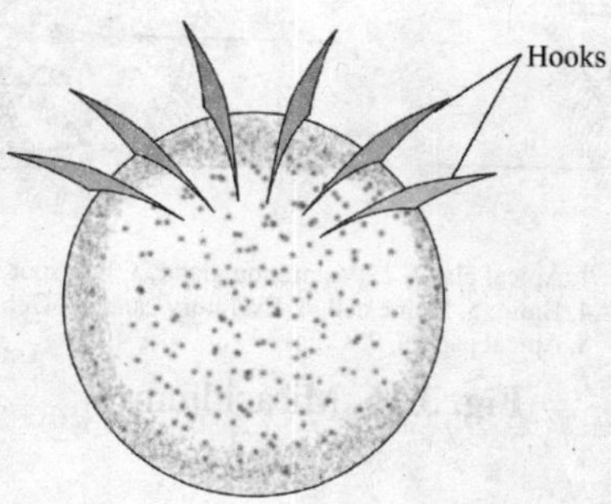

(a)

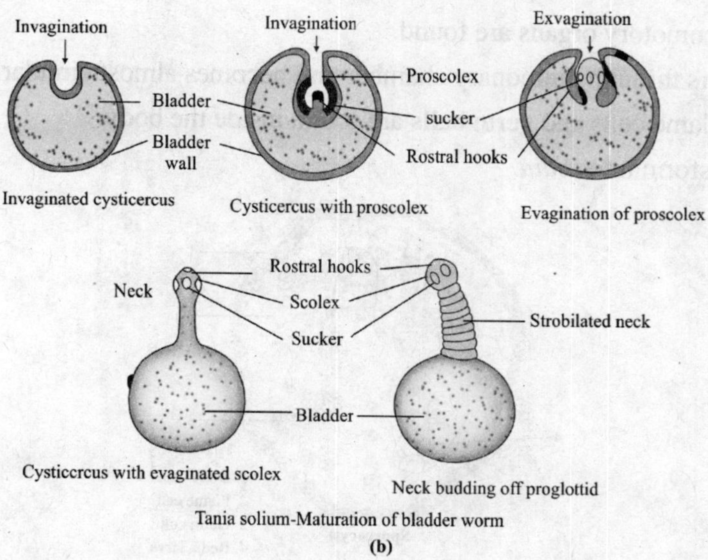

Tania solium-Maturation of bladder worm

(b)

Fig. 3.15. a. Hexacanth, b. Cysticercus and maturation of bladder worm.

Miracidium (Size - 0.13 mm)

1. It is first larval stage of *Fasciola hepatica* that is *free swimming* in habit.
2. Body covered with cilia.
3. Body triangular in shape.
4. Apical papilla on anterior end. It can penetrate the alimentary canal.
5. A pair of eye spot located behind papilla.
6. A pair of flame cells, parenchyma and germ cells are found in the body.
7. It is not survived for long time. If it find its secondary host, *Lymnaea trancatula* (snail) within 8 hours, it developed further otherwise dies. It transform into *sporocyst.*

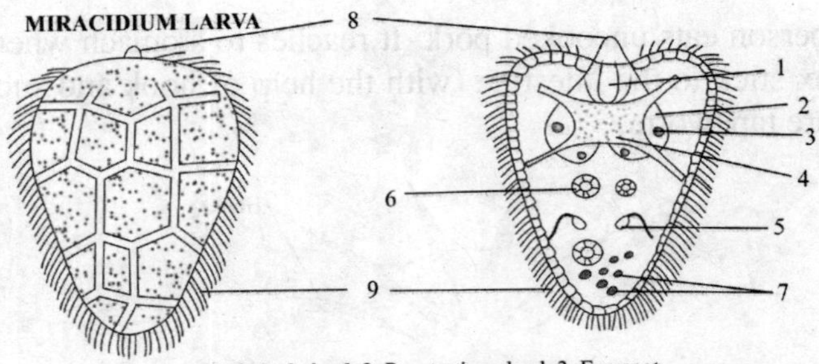

Fig. 3.16. Miracidium

Sporocyst (Size - 0.01 mm)

1. It is second larval stage of *Fasciola hepatica.*
2. No locomotory organs are found.
3. It enters through pulmonary chamber and becomes almost circular.
4. Four flame cells and germ balls are found inside the body.
5. It transform into *redia.*

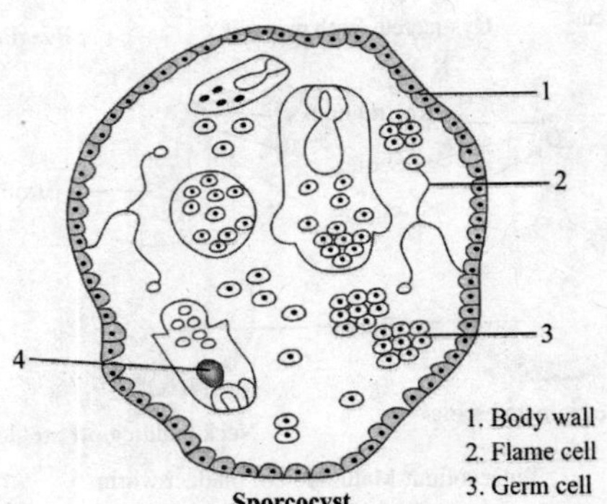

Fig. 3.17. Sporocyst

Redia (Size 1.3-1.51 mm)

1. It is developed inside sporocyst while in pulmonary chamber.
2. Body elongated that bears mouth at anterior end. It bears muscular pharynx.
3. Below pharynx is *collar*. Birth pore is situated near to it.
4. Cercaria and germ cells are found inside the body.
5. It is matured within 12-18 days.

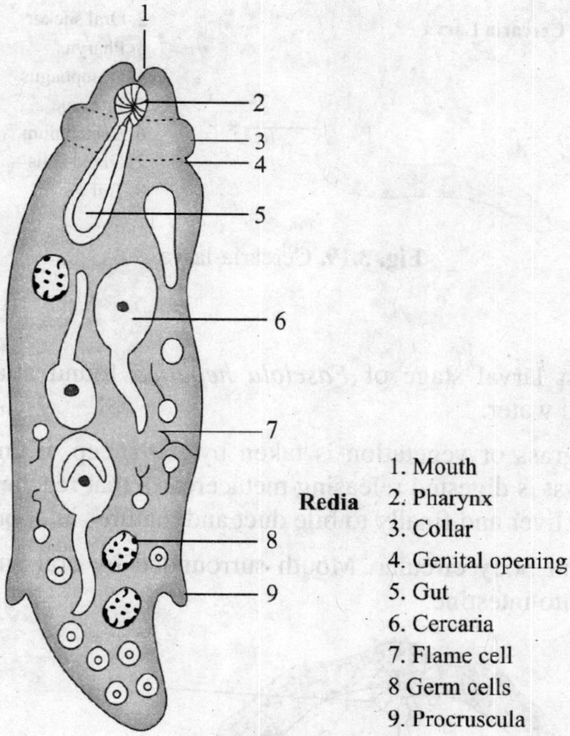

Fig. 3.18. Redia

Cercaria (Size 0.6 mm)

1. It is penultimate larval stage that got developed inside redia. Cercaria is developed in 20 days old redia.
2. It comes out through birth pore and from the body of snail. Body ovoid with a distinct tail.
3. Body heart shaped that bears *oval* and *ventral sucker*. Mouth is located in the middle of oral sucker. It leads into pharynx.
4. It swims about 2 hours. Tail disappears and secrete a cyst. Encysted larva is called *metacercaria* that stick to some vegetation in the water.

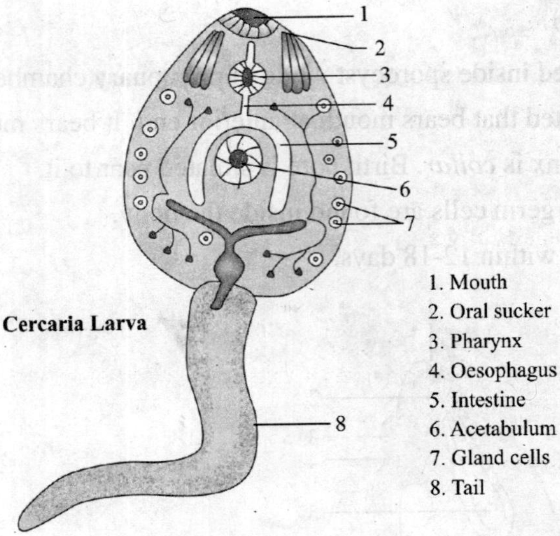

Fig. 3.19. Cercaria-larva

Metacercaria

1. It is the last larval stage of *Fasciola hepatica* found attached to some grass or vegetation in water.
2. When this grass or vegetation is taken by the sheep, it enters into the alimentary canal and cyst is digested releasing metacercaria that reaches to intestne. On 4th day it reaches to liver and finally to bile duct and matures in about 4 weeks.
3. Tail is absent body circular. Mouth surrounded by oral sucker. Pharynx is forked that opens into intestine.

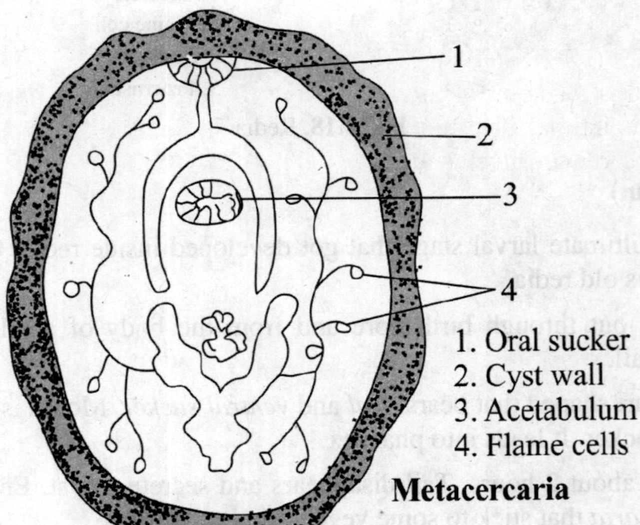

Fig. 3.20. Metacercaria

Leech : Testicular nephridia

1. It is horse-shoe shaped and consists of 6 parts viz., main lobe, vesicle, apical lobe, initial lobe, inner lobe and ciliated organ. From segment 6 to 11 are pre testicular nephridia. Rest are testicular.
2. Main lobe has two limbs, shorter and longer
3. Apical lobe is slightly swollen at anterior end.
4. Coiled around apical lobe is initial lobe.
5. Inner lobe extends between anterior and posterior limb of main lobe.
6. Ciliated organ hangs to inner wall with 4 or 5 trabeculae/strands.
7. It serves the function of excretion.

In embryo it has connection with nephridium but looses it in adult.

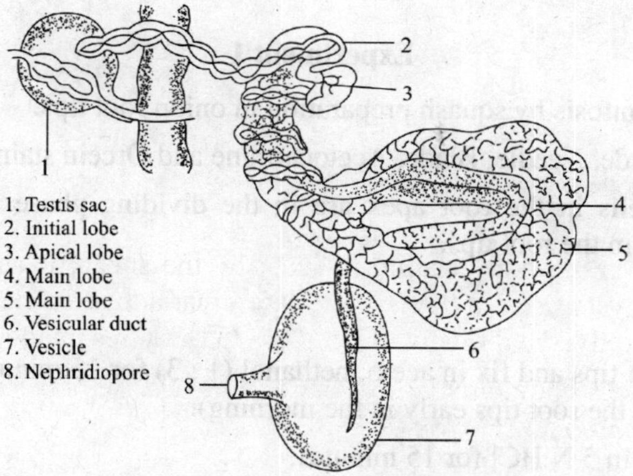

Fig. 3.21. Leech : Testicular nephridia

The tracheal wall is made up of three layers viz. basement membrane, epithelium and inner cuticle. Inner cuticle is intima. The intima is produced into tenidia (spiral thickening). It protects it from collapsing and remain it always open. They carry air for respiration and major part of respiratory system.

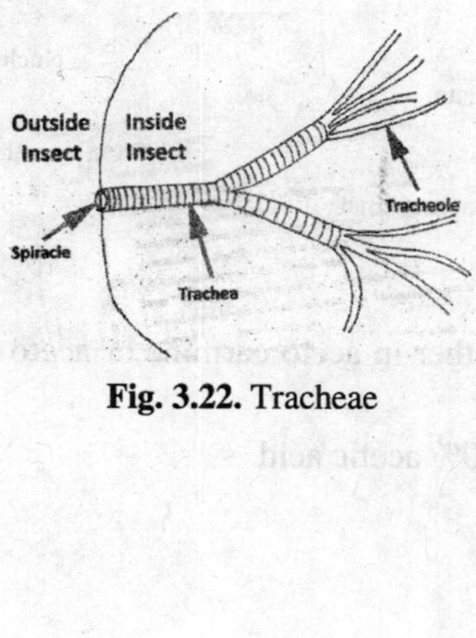

Fig. 3.22. Tracheae

Unit - 4
Exercises in Cell Biology

Experiment 1

Object - To study mitosis by squash preparation in onion root tips.

Requirements - Slide, Needle, Blade, Acetocarmine and Orcein stain. Aceto- methanol etc.

Principle - The cells in the root apex are in the dividing phase. Mitotic cells divisions therefore can be studied in the root tips.

Procedure

1. Cut the root tips and fix in aceto methanol (1 : 3) for 15 minutes. (Note : Cut the root tips early in the morning).
2. Keep them in 5 N HCl for 15 minutes.
3. Pass the root tips in 95% alcohol 3 changes of 5 minutes each.

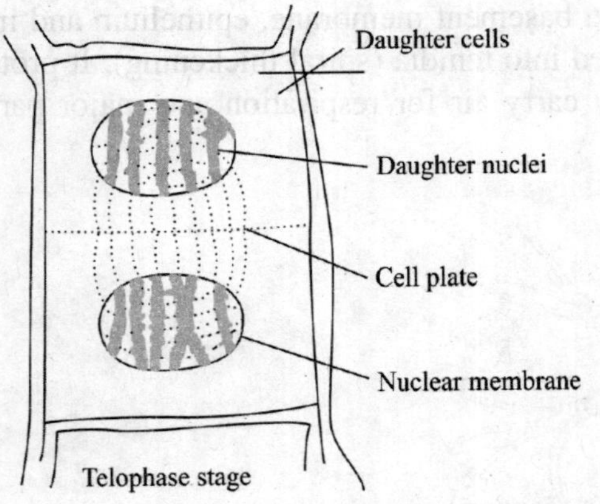

Telophase stage

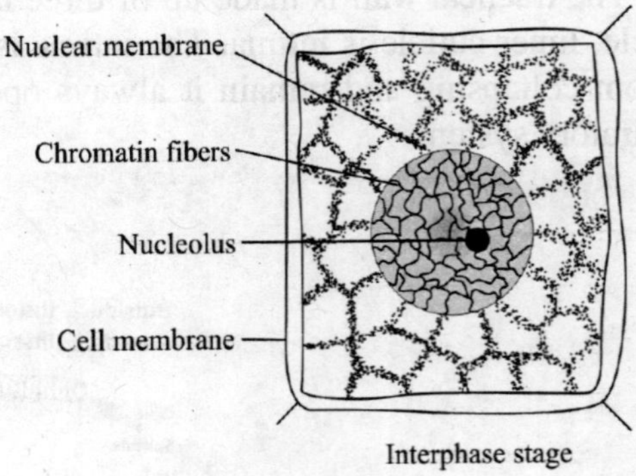

Interphase stage

4. Now stain the tips either in aceto carmine or aceto orcein or both 1 : 1 for about 20 minutes.
5. Wash the tips with 50% acetic acid.

6. Keep on a clean slide in a drop of 50% acetic acid and cover with cover slip. Squash gently with the help of little pressure of thumb and seal with nail polish.

Result - Different stages of mitosis are seen. Following stages are visible - Prophase, Metaphase, Anaphase and Telophase. Interphase nuclei are also seen.

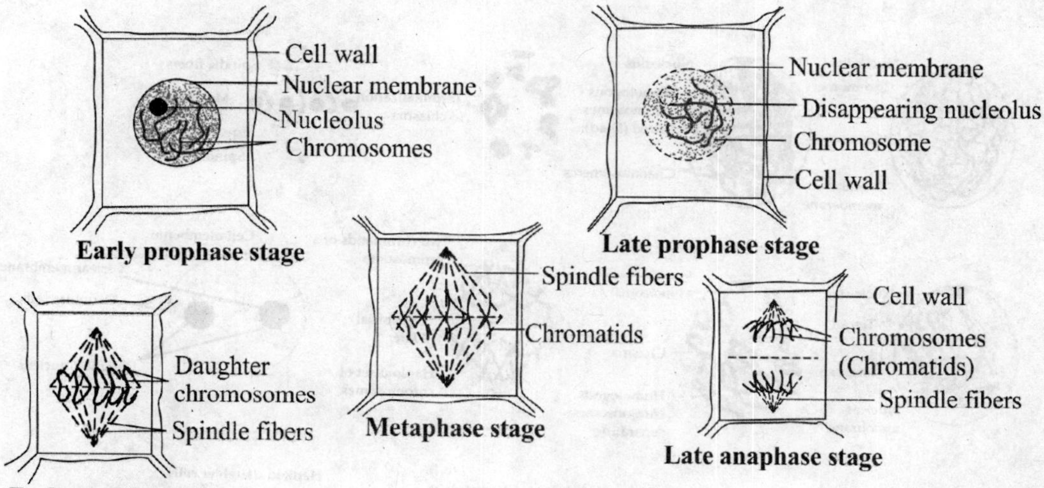

Fig. 4.1: Various stages of mitosis

Second Method

Procedure

1. Cut the root tips and keep on the slide in a drop of Allen's fluid (saturated Picric acid - aquous 75 ml, Formalin 15 ml, Acetic acid 10 ml and Urea 1 gm) for 30-40 seconds.
2. Add 1 drop of acid alcohol (Conc HCl 100 ml and Isopropyl alcohol 200 ml).
3. Add 2% aceto orcein (few drops)
4. Cover with coverslip and gently squash.

Results - Chromosome are well spread and all stages of mitosis are visible.

Experiment 2

Object - To study meiosis in grasshopper testes.

Requirements - Living Grasshopper (male) or Cockroach (male) Insect Ringer's solution or physiological saline, acetocarmine or acetoorecin, slide, coverslips, microscope, aceto methanol (1 part acetic acid + 3 part alcohol).

Principle : Acetic acid causes the tubules to soften so that they are easily squashed and broken, as a result of which the different stages of meiotic division are easily observed.

Procedures : Dissect out the testes from the grasshopper which appear orange or yellow in the abdominal segments (4, 5 and 6 segments). Put them in insect Ringer's solution and afterwards fix in aceto methanol by keeping them into this solution for about 5 minutes. Now put them on a clean slide and pour 2-3 drops of stain. Allow to take the stain for 8-10 minutes. Put a coverslip on it and press gently with the help of right hand thumb. This is called squash preparation. Seal the margins of coverslip with nail polish and observe the slide under microscope.

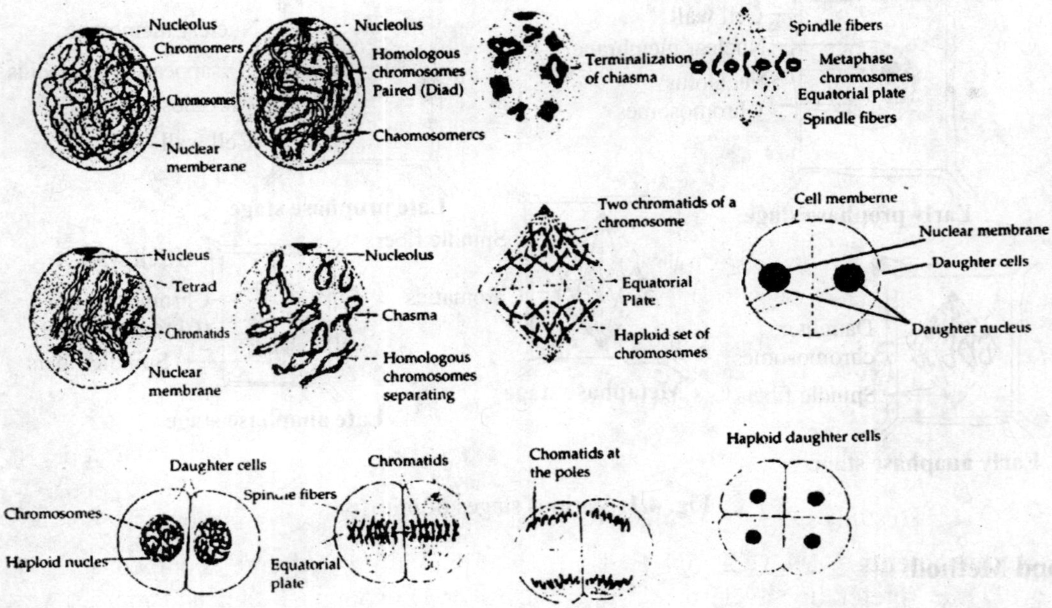

Fig. 4.2 : Various stages of meiosis

Result - Different stages of meiotic division as **leptotene, zygotene, pachytene, diplotene** etc. are seen.

Experiment 3

Object - To study the giant chromosomes in the salivary glands of Drosophila.

Requirements : Drosophila larvae (third instar) or Chironoums larvae (mature), aceto carmine, aceto orcein, Insect Ringer's solution or physiological saline, forceps and needle, slide, coverslip, nail polish, spirit lamp, blotting paper etc.

Procedure : Dissect out salivary glands of 3rd instar larvae in insect Ringer's solution or physiological saline. As you pull out the mandibles, salivary glands also comes along them. Separate them and fix them in 2-3 drops of acetomethanol. Stain the salivary glands either is aeotocarmine or aceto orcein for 8-10 minutes. Put 1-2 drop of acetic acid on a clean slide and transfer the stained salivary glands into it. Put the coverslip and press gently. Seal the margins of coverslip with sealing wax or nail polish and observe under the microscope.

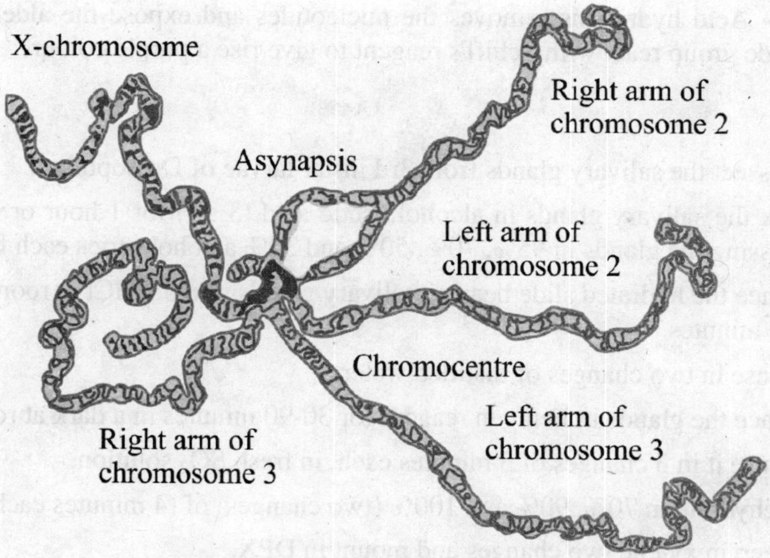

Fig. 4.3: Polytene Chromosomes of Drosophila

Experiment - 4

Object - To study DNA by Feulgen reaction in salivary gland chromosomes.

Requirements - 1N HCl, 5N HCl, 10% Sodium metabisulphite, Feulgen stain, alcohol series. Tissue (preferably liver) or salivary glands of Drosophila Pestel and mortar, graduated tubes, centrifuge machine, distilled water, etc.

Feulgen Reagent

Take 48 ml of 1N HCl and mix it into 16 ml of double distilled water. Warm it at 60ºC.

Add to it, 0.8 gm Fuchsin slowly.

Boil it (slowly) for 3 minutes.

Cool it (60ºC) and filter. Allow to cool upto 30ºC and then add 0.32 gm sodium metabsulphite. Stir, keep the bottle in dark for about 12 to 24 hrs. The colour is straw yellow. Now add 0.5 gm activated charcoal. Keep it for 12 hours in a cool place and filter through whatman filter paper. Keep in dark glass bottles in a refrigerator.

Modification - 0.5 gm basic funchsin is dissolved in 100 ml of boiling distilled water and solution is cooled upto 58ºC and filtered.

Temperature is allowed to fall upto 26ºC and add to it, 10 ml of 1N HCl + 0.5 gm $K_2S_2O_5$ (Potassium metabisulphite). Keep the solution in a conical flask wrapped with black paper for overnight (24 hrs). Magenta coloured solution turn straw colour (due to bleaching action of SO_2 released by the action of HCl on potassium metabisulphite). Add a bit of activated charcol and keep it in a dark cool place and filter.

SO_2 solution - Take 20 ml 10% sodium metabisulphite and 20 ml 1N HCl. Add 360 ml of distilled water to it.

Principle - Acid hydrolysis removes the nucleotides and expose the aldehyde group. The exposed aldehyde group react with Schiff's reagent to give rise a purple colour.

Procedure

1. Dissect the salivary glands from 3rd instar larvae of Drosophila.
2. Fix the salivary glands in alcohol acetic acid (3 : 1) for 1 hour or so. Hydrate it by passing the glands in 95%, 70%, 50% and 30% alcohol series each for 15 minutes.
3. Place the hydrated slide bearing salivary glands into 5 NHCl at room temperature for 20 minutes.
4. Rinse in two changes of distilled water.
5. Place the glands in Feulgen reagent for 30-90 minutes in a dark at room temperature.
6. Rinse it in 3 changes of 3 minutes each, in fresh SO_2 solution.
7. Dehydrate in 70%, 90% and 100% (two changes) of (4 minutes each).
8. Keep in xylene two changes and mount in DPX.

Result - DNA appears in shades of reddish purple.

(The intensity of the colour is proportional to the amount of DNA).

SECOND METHOD

Materials

1N HCl

Schiff's reagent

Sulphite wash

Fast green

(For preparation of reagents see appendix 1)

Procedure

(1) Fix the salivary glands of **Drosophila** or **Chironomus** or any other tissues (testes, ovary etc.) in carnoy's fixative for at least 20 minutes.

(2) Rinse in absolute alcohol for 3-5 minutes.

(3) Hydrate upto 70% alcohol.

(4) Rinse in distilled water.

(5) Incubate in 1N HCl at 60ºC temp. for 10 minutes.

(6) Again rinse in distilled water.

(7) Stain in Schiff's reagent for 10 minutes.

(8) Give 3 changes of sulphite wash of 2 minutes each.

(9) Wash in running tap water for 15 minutes (till sulphite smell is gone)

(10) Rinse in 70% alcohol.

(11) Counter stain in fast green solution for 4-5 seconds.

(12) Dehydrate in absolute alcohol.

(13) Clear in xylene.

(14) Mount in DPX

Result - DNA appears as purple to reddish purple.

THIRD METHOD

Feulgen reaction is used for quantitative measurement of DNA. In Feulgen staining the nuclei are selectively stained and the intensity of the colour is directionally proportional to the amount of DNA present.

Materials required

Fixative - Alcohol acetic acid (3 : 1)

Bleaching solution - 10% Potassium metabisulphite + 5 ml HCl + 500 ml DW.

Tissue - Any insect tissue preferably testis and ovary.

Schiff's reagent - Take 0.5 gm. basic fuchsin in a conical Flask. Add to it 100 ml boiling water.

Mix thoroughly and allow to cool to 50ºC. Add 0.5 gm potassium metabisulphite and 10 ml 1N HCl.

(1N HCl = 10 ml conc. HCl + 90 ml. DW)

keep in air tight bottle for a day. Add 0.25 gm activated charcoal and filter it. If it is not colourless further add activated charcoal and keep in refrigerator.

Procedure

1. Fix the tissue in alcohol acetic acid for 3-10 hours.
2. Transfer to 70% alcohol. Dehydrate in 90% alcohol, absolute to xylene.
3. Embed the tissue in wax (paraffin)
4. Make blocks and cut the section at microtome (5.8 μ thick).
5. Remove the wax (deparaffinise) from the tissue by passing through descending series of alcohol.
6. Bring the slide to 1N Hcl for 5-6 minutes at 60ºC temperature.
7. Bring it to 1N HCl room temperature.
8. Rinse in distilled water.
9. Stains in Schiff's reagent for 1-2 hrs.
10. Drain and transfer through 3 changes of bleaching solution (5 ml 15 of pot. metabi sulphite + 5 ml 1 N HCl + 90 ml distilled water) for 1 minute each.

11. Rinse in distilled water.
12. Dehydrate upto 90% alcohol.
13. Stain in 0.005% Fast green in 95% alcohol.
14. Dehydrate in absolute alcohol. Clear in xylene and mount in DPX

Result - The nuclei stain purple representing the DNA.

Experiment 5

Object - To study cell permeability using mammalian RBCs.

Requirements - Normal saline, Hypotonic solution, Hypertonic solution, disposable needle, sprit, slide, coverslip, compound microscope.

Principle - When RBCs are bathed in hypotonic solution, they tend to increase in size as the solute molecules (H_2O) are in higher concentration in outside medium as compared to internal contents of RBCs. As a result they move inside the cell and it increases in size. On contrary when kept in hypertonic solution, the water from within the cell comes out and the cell shrinks. It is called *crenation.* In hypotonic solution sometimes the cell burst out releasing haemoglobin, it is called *haemolysis.* There is no effect or RBCs in isotonic solution.

Procedure

1. Sterilize your left hand ring finger and puncture it with the help of disposable needle to ooze-out a drop of blood.
2. Place this drop on a clear slide at three places.
3. Put normal saline, hypotonic solution and hypertonic solution to each of it.
4. Observe the cells after 5-7 minutes.

Results

The RBCs in hypertonic solution got shrinks and in hypotonic solution swollen. No effect in isotonic solution.

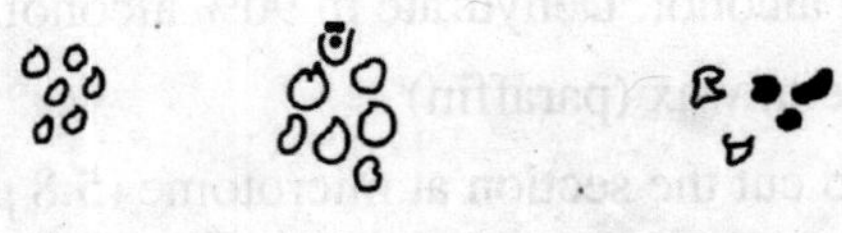

Normal size cells (Isotonic solution) — Swollen cells (hypotonic solution) — Shrink cells (hypertonic solution)

Fig. 4.4.

Unit - 5
Exercises in Genetics

1. Life Cycle of Drosophila

Drosophila is commonly called as *fruit fly* has life cycle of very short duration. The life cycle takes about 11 days to complete. The average life span of male fly is 32-33 days and 26-27 days of females.

Artificial Diet

(i) Agar agar or Gelatin	-	10 gm
		(Dissolve agar-agar in warm 300 ml) distilled water)
(ii) Non-crystaline sugar	-	65 gm
(iii) Cornflour	-	75 gm
(iv) Yeast powder	-	10 gm

Dissolve the above ingredients in 450 ml distilled water and heat the contents. Add 2 ml *propionic acid.* Transfer this semi liquid paste into milk bottles/ culture bottle and allow to cool at room temperature. Wipe off the water vapour (condensed vapours) by blotting paper & use as artificial diet.

Procedure

1. Release 10 adult flies (sex ratio 1 : 1) in a culture bottle and plug the mouth of the bottle with the help of cotton.
2. The flies mate in the bottle and lay eggs in the diet. Remove the flies from the culture bottle after 2-3 days.
3. Now trace the life cycle which gets completed in the following stages.

Eggs - The eggs are small white in colour provided with two anchors. They hatch within 24 hours.

Larva - The eggs are hatched into 1st instar larva which is very small creamish in colour. It metamorphose into IInd instar after a day. The IInd instar larva after about two days change into IIIrd instar which is voracious eater and crawl actively within the food. The tunnels and streaks caused by crawling can be easily seen by the side of the culture bottle. After about 3 days it metamorphose into *pupa*.

Pupa - 3rd instar larvae attach itself to the bottle and becomes inactive, do not feed. It change into pupa which is light brown in colour. After about 4 days adults flies comes out of it.

Adult - The adult is of 2 mm in size and a little brown fly.

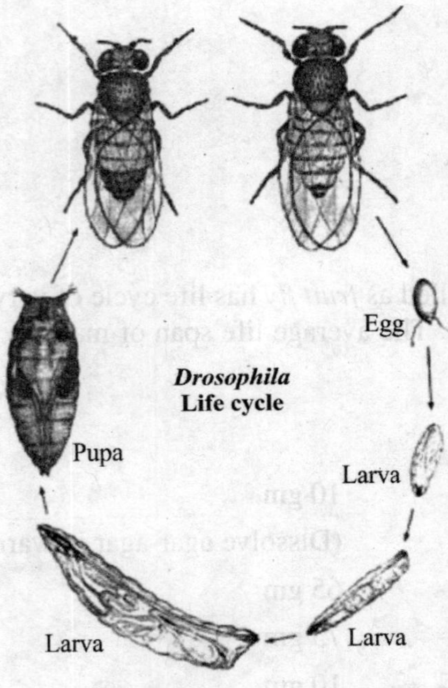

Fig. 5.1. Life stages of Drosophila

2. Identification of Male and Female Drosophila

Narcotise a few adult flies with the help of ether vapours or liquid CO_2 and examine on the basis of following characteristics

S.No.	Character	Male	Female
1	Size	Smaller than female	Larger than male.
2	Abdomen	Tip of the abdomen rounded and black largely	Tip of the abdomen pointed and less black
3	Abdominal segments	5 segments	7 segments
4	Sex comb	Present on the ***tarsus*** of the 1st pair of legs	Absent
5	External genitelia	Claspers and penis present	Absent
6	Life span	Comparatively more than female	Comparatively less than male

Fig. 5.2. Adult fly

3. Identification of Wild and Mutant Types

Nacrotize few flies from the culture and observe the following characters -

Eye colour (Red or White)

Wings (Developed or Vestigial)

Body Colour (Brown, Yellow or Sooty or Black).

Any bar in the eyes.

S.No.	Character	Wild	Mutant
1	Eye colour	Red	White
2	Wings	Developed	Vestigial
3	Body colour	Brown	Ebony or Black or Sooty or sepia

Note : White eye mutation on *x-chromosome,* vestigial wings on chromosome-2 represented by (vg) and Black body colour (ebony) on chromosome no. 3, represented by (e).

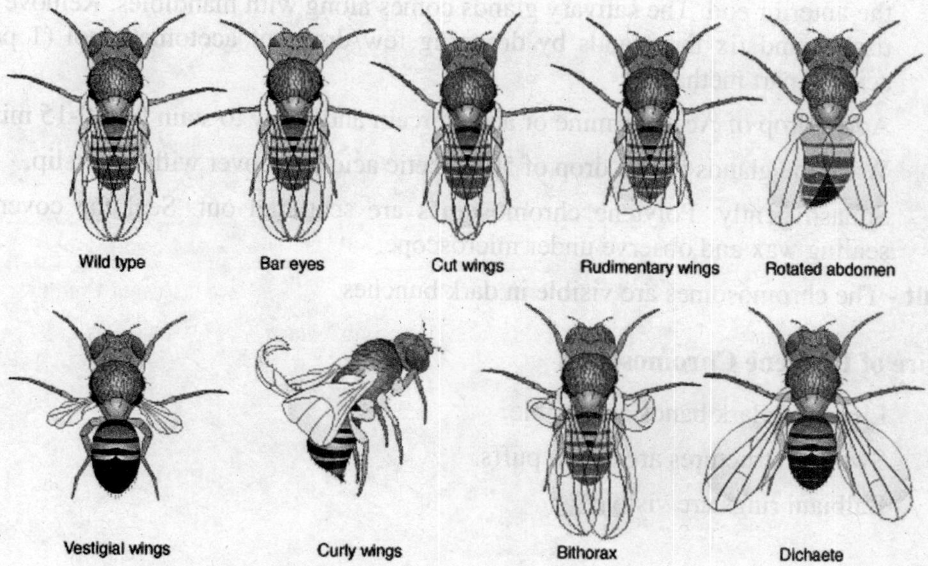

Fig. 5.3. Mutants of Drosophila

4. Slide Preparation of Salivary gland Chromosomes

Requirements

Ringer's solution, Aceto-carmine or Aceto orcein stain, Aceto-methanol (fixative). Fine dissecting instruments like needle, forceps, scissors, blade, brush etc.

Ringer's Solution

NaCl	-	0.65 gm
Na_2CO_3	-	0.62 gm
$CaCl_2$	-	0.03 gm
Kl	-	0.025 gm
Distilled water	-	100 ml

(Dissolved the contents)

Dissolve 2 gm carmine powder in 100 ml 45% boiling acetic acid. Again heat for 10 minutes and cool and filter.

Aceto orcein - Dissolve 2 gm orcein powder in 100 ml 45% boiling acetic acid. Again heat for 10 minutes cool and filter it.

Procedure

1. Place a drop of Ringer's solution on slide and transfer a 3rd instar larvae into it.
2. Identify the anterior end of the body. Mandibles are located at this end.
3. Hold the rear part of the body with the help of a forceps and pull out the contents of the anterior end. The salivary glands comes along with mandibles. Remove the other tissues and fix the glands by dropping few drops of acetomethanol (1 part acetic acid; 3 part methanol).
4. Add a drop of Acetocarmine or aceto orcein and allow to stain for 10-15 minutes.
5. Wash the glands with a drop of 50% acetic acid and cover with coverslip.
6. Squash gently. Polytene chromosomes are scattered out. Seal the coverslip with sealing wax and observe under microscope.

Result - The chromosomes are visible in dark bunches.

5. Structure of Polytene Chromosomes

1. Light and dark bands are visible.
2. Swollen structures are called puffs.
3. Balbiani rings are visible.

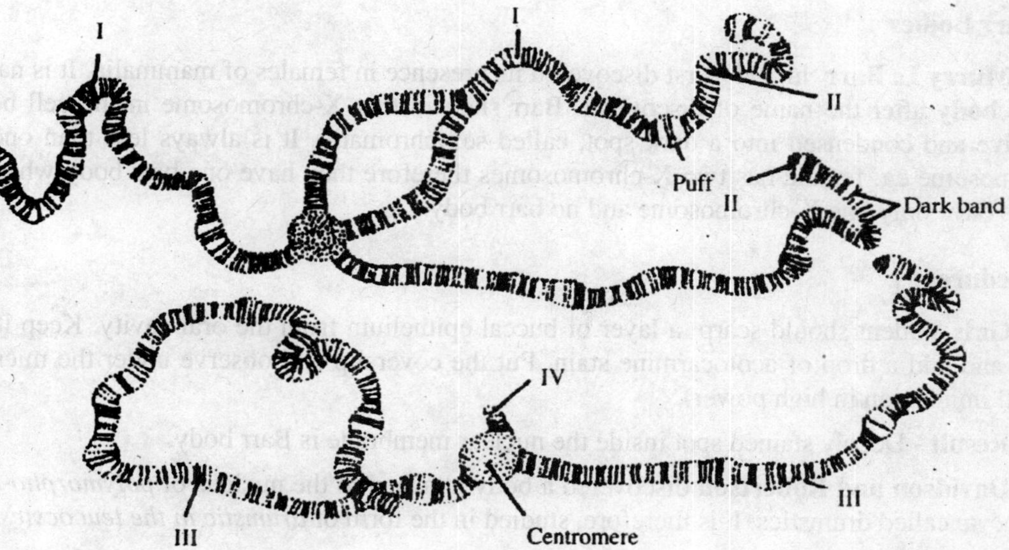

Fig. 5.4. Polytene chromosome

6. Sex Comb

1. It is located at the tarsi of forelegs of male.
2. It is comb like structure containing 10 teeth.

Procedure

Narcotize the male fly and observe under dissecting microscope. Remove the 1st pair of legs from the body and mount in glycerine. For permanent slide dehydrate the slide in alcohol series and mount in DPX.

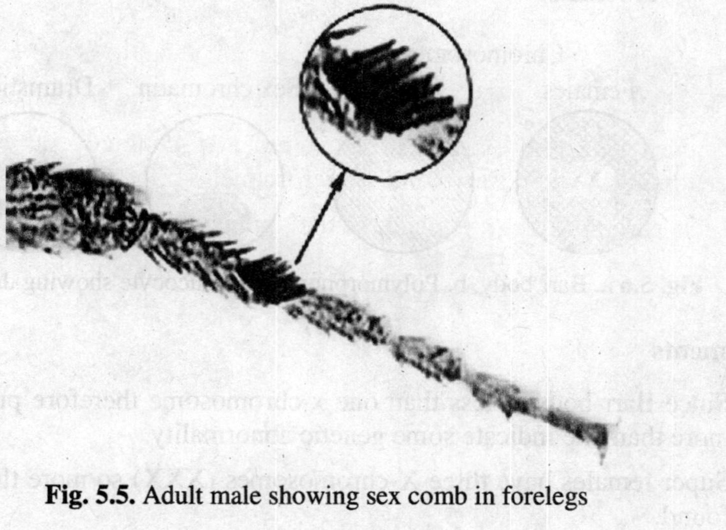

Fig. 5.5. Adult male showing sex comb in forelegs

7. Barr Bodies

Murry L. Barr in 1949 first discovered its presence in females of mammalia. It is named as **Barr body** after the name of discoverer. Barr showed that X-chromosome in the cell becomes inactive and condensed into a dark spot, called sex chromatin. It is always less than one of X-chromosome eg. female has two X-chromosomes therefore they have one barr body whereas the males have only one X-chromosome and no barr body.

Procedure

Girls student should scarp a layer of buccal epithelium from the oral cavity. Keep it on the slide and add a drop of acotocarmine stain. Put the coverslip and observe under the microscope (in oil immersion in high power).

Result - Deeply stained spot inside the nuclear membrane is Barr body.

Davidson and Robertson discovered a body attached to the nucleus of *polymorpho-nuclear leucocyte* called drumstics. It is therefore, studied in the form of *drumstic in the leucocytes.*

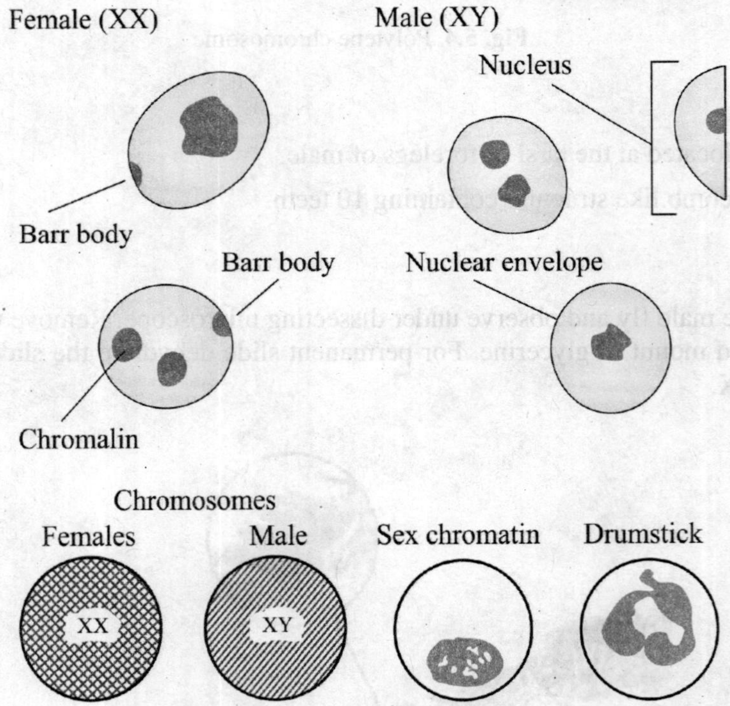

Fig. 5.6 a. Barr body, b. Polymorphonuclear leucocyte showing drumstics

Special Comments

1. Since Barr body is less than one x-chromosome therefore presence of a Barr body more than one indicate some genetic abnormality.

   Super females have three X-chromosomes (XXX) so more than one barr bodies are found.

2. The genotype of *Klinefelter's* syndrome is XXY therefore, one Barr body is found.
3. The genotype of *Turner's syndrome* is XO therefore, no Barr body is found in such persons.
4. They can be seen without staining.

8. Human Chromosomes

Tigio and Levin (1956) stated the presence of 46 chromosomes in man (2n = 46). Every body cells have 46 chromosomes, out of which 22 are identical in both male and female called *autosomes* or *somatic* whereas 23rd is different called *sex chromosome*.

22 + XX ♀ (woman)
22 + XY ♂ (man)

Y chromosome is smaller than X. When the chromosomes are arranged according to shape and size for cytological studies. It is called *Karyotype.*

Pictorial representation of karyotype is *Idiogram.*

Human karyotype is briefly discussed here.

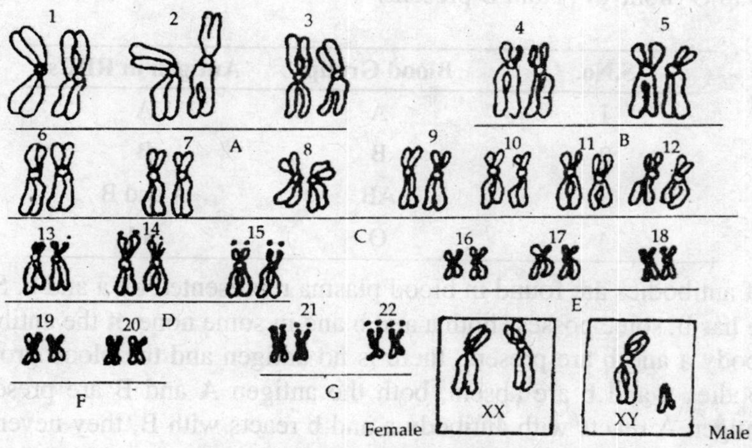

Fig. 5.7: Human Idiogram.

Autosomes are numbered on the basis of shape, size and position of centromere, 22 pair of chromosomes are arranged in seven different groups (A to G).

Group A - 3 pair (1 to 3), large metacentric chromosomes.

Group B - (4 & 5), large (comparatively smaller than group A) submetacentric chromosomes.

Group C - Medium size (6 to 12), submetacentric chromosome.

Group D - Medium size (13 to 15) acrocentric chromosomes.

Group E - Comparatively smaller than group D, metacentric (16 to 18)

Group F - Small size (number 19 to 20) metacentric chromosome.

Group G - Smallest in size (number 21 to 22) metacentric chromosomes.

Sex Chromosomes

X and Y chromosomes resembles in shape and size with the chromosomes of C Group, Y chromosome is smaller and resembles with G group chromosomes. The sex chromosomes are indicated in last.

9. Blood Groups

Karl Landsteiner (1900) propounded the presence of two *antigens* A and B in human RBCs. On the basis of this discovery the human blood groups were divided into three groups viz. A, B, O. Later in the year 1902 **Decastello** and **Sturil** discovered the 4th group -AB Antigens are found in human red blood cells and antibodies in the plasma. If these antigens and antibodies are brought together, they agglutinate. Some persons possess antigen A, some possess B, some has both A and B and some has no antigens. On this basis four blood groups have been identified.

(i) Group A (antigen A present)

(ii) Group B (antigen B present)

(iii) Group AB (both antigen A and B present)

(iv) Group O (none of A and B present)

S.No.	Blood Group	Antigen in RBCs
1	A	A
2	B	B
3	AB	A and B
4	O	Nil

Two type of antibodies are found in blood plasma represented by a and b. Some people have a antibody, some has b, some possess both a and b and in some none of the antibodies are present. When both antibody a and b are present, there is no antigen and the blood group is O. Whereas when both antibodies a and b are absent, both the antigen A and B are present and the blood group is AB. Antigen A reacts with antibody a and b reacts with B, they never present together. For a given antigen an opposite antibody is present in the man eg. a person with antigen A possess antibody b and with antigen B has antibody a. This is only the reason why blood in the human body never agglutinate.

S.No.	Blood Group	Antibody (in plasma)	Antigen (in RBCs)
1	A	b	A
2	B	a	B
3	AB	Nil	A and B
4	O	a and b	Nil

A person with O group can donate blood to anyone hence called **Universal donor** because not antigen is present in the RBCs, persons with Group AB can receive blood from anyone hence called **Universal recipient** because there is no antibody present in their plasma.

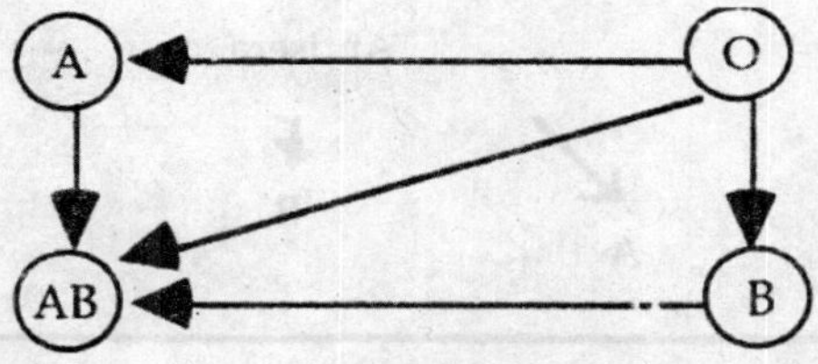

– A can donate to AB can receive from O and A groups.

= B can donate to AB and can receive from O and B.

≡ AB cannot donate to any other but can receive from O, A and B groups (hence **universal recipient**). It can donate to AB only

≅ O can receive from O group only but can donate to anyone (hence **Universal donor**)

Object - To identify the blood groups.

Requirements - Anti-sera A, B, AB needle, sprit lamp, alcohol, slide and microscope.

Principle

When the blood is mixed with a given antisera, two types of reaction takes place

(i) agglutination

(ii) no agglutination

With the antisera, agglutination occurs, indicate the blood group eg. if the agglutination occurs with antisera A then the blood group will be A, if the agglutination occurs with anti sera B then blood group is B and if agglutination takes place by both antisera A and B, the group is AB, O blood group is called universal donor and AB as universal recipient.

Procedure

(i) Rapid slide test

(ii) Test tube method.

Rapid Slide Test

(i) Place a drop of blood at three selected places on a slide and add one drop saline to each drop of blood.

(ii) Add one drop each of antisera A, antisera B and antisera AB.

(iii) Mix with a glass rod and note if there is any agglutination.

Result - Agglutination of RBCs with the antisera indicate positive reaction (*Positive test)* and no agglutination indicate *Negative test.*

S.No.	anti-A	anti-B	anti-AB	Blood Group
1	-	+	+	B
2	+	+	+	AB
3	+	-	+	A
4	-	-	-	O

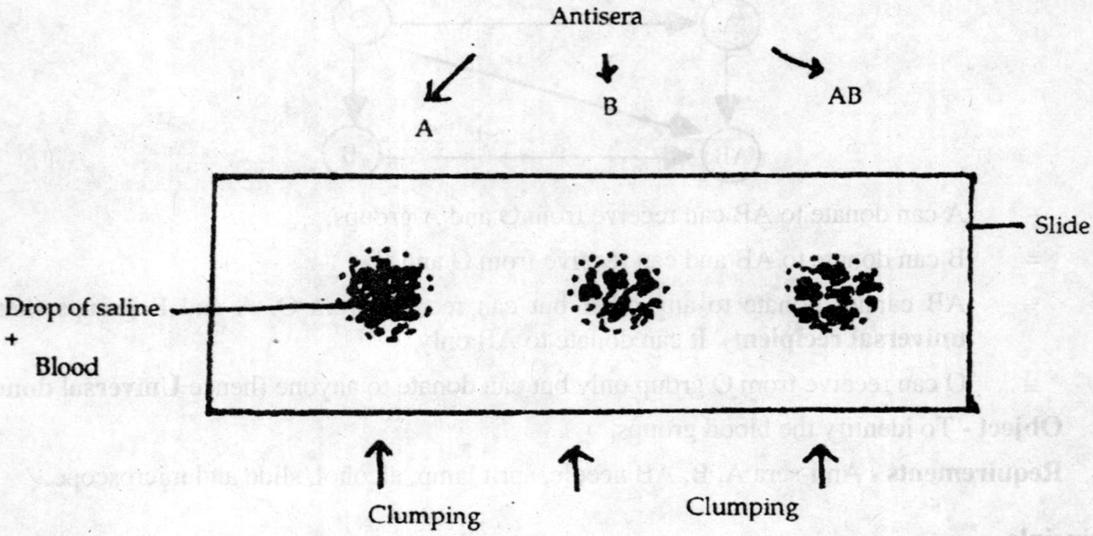

Fig. 5.8. Clumping of RBCs with antisera A, B and AB.

Test Tube Method

1. Take 3 test tubes and mark A, B and AB.
2. Take 4% suspension of the blood in normal saline and add in each tube.
3. Add antisera A in A tube, antisera B in B tube and antisera AB in AB tube. Note the agglutination and results in your record.
   (It can be observed under microscope also)

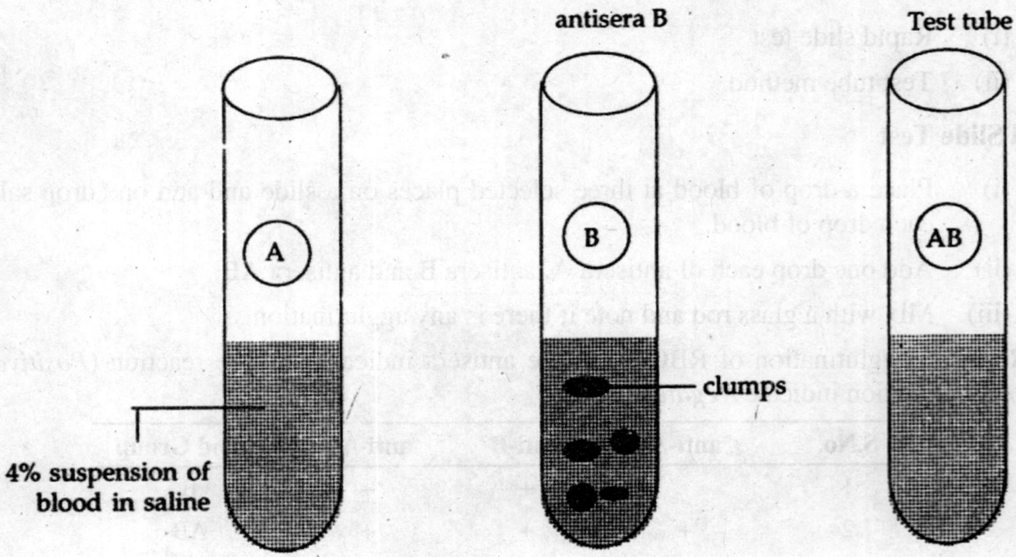

Inference - Since the agglutination takes place with antisera B therefore the group is B.

To identify Rh Blood Group

It is like rapid slide test. Use antiser. If the agglutination takes place with antiser, the group is Rh positive otherwise Rh-negative. 99% of the human population is Rh positive.

If your blood agglutinates in	You are blood type
Anti - A, but not Anti - B	A
Anti - B, but not Anti - A	B
Anti - A and Anti - B	AB
Neither Anti A or Anti - B	O
Anti - D	Rh positive (Rh^+)

10. Numerical Problem based on Mendalian Laws

Female (♀) Male (♂)

Normal Vestigial Wing

$v_g^+ v_g^+$ × $v_g v_g$

v_g^+ vg F1

$v_g^+ + v_g$ $v_g^+ + v_g$

(Normal Female) (Normal Male)

♀ \ ♂	v_g^+	v_g
v_g^+	$v_g^+ v_g^+$ Normal	$v_g^+ v_g$ Normal
v_g	$v_g^+ v_g$ Vestigial	$v_g v_g$ Vestigial

3: 1 = Phenotype Normal (3) : Vestigial (1)

1: 2: 1 = Genotype

2. If one homozygous male rabbit is crossed with homozygous female rabbit (white) then what is the phenotypic ratio of F_1 generation. If F_1 generation rabbits are interbreeded together then what is the genotypic ratio of F_2 generation. (Given : Black is dominant over white colour).

Allele for Black colour `B'

Allele for Whit colour `b'

Male (♂) (Black) × Female (♀) (White)

BB bb

↓ ↓

B b – (Generation)

↓

Bb ....F_1 (Generation)

(All Black)

♀ \ ♂	B	b
B	BB (Black)	Bb (Black)
b	bB (Black)	bb (White)

3 : 1 = Phenotype (Black : White)

1 : 2 : 1 = Genotype

BB (Pure Black) Bb (Hybrid black) bb (White)

3. Demonstration of test cross in Drosophila

Female (Red eye) $\overset{+}{w}$ ϕ ϕ w   **Male (White eye)** w ϕ ɿ

♀ \ ♂	w ϕ	ɿ
$\overset{+}{w}$ ϕ	$\overset{+}{w}$ ϕ ϕ w Female (Red eye)	$\overset{+}{w}$ ϕ ɿ Male (Red eye)
w ϕ	w ϕ ϕ w Female (White eye)	w ϕ ɿ Male (White eye)

Phenotypic Ratio : 50% Red eye : 50% white eye

Genotypic Ratio : 1 : 1 : 1 : 1

Red eye : Red eye : White eye : White eye

♀ ♂ ♀ ♂

4. **If carrier white eyed female drosophila is crossed with red eyed male then what is the ratio of eye colour?**

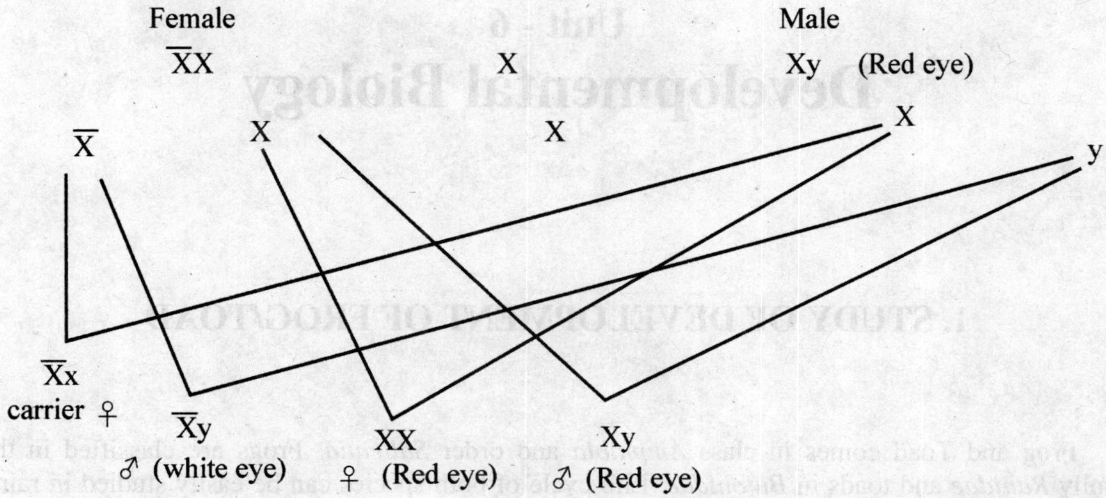

Ratio

25%	=	Carrier female
25%	=	Red eyed female
25%	=	White eyed male
25%	=	Red eyed Male

5. **If a colour blind person marry a normal female then what is the percentage of colour blindness in their childrens? Colour blindness is a sex-linked disease. Genes of colour blindness are located on chromosome X.**

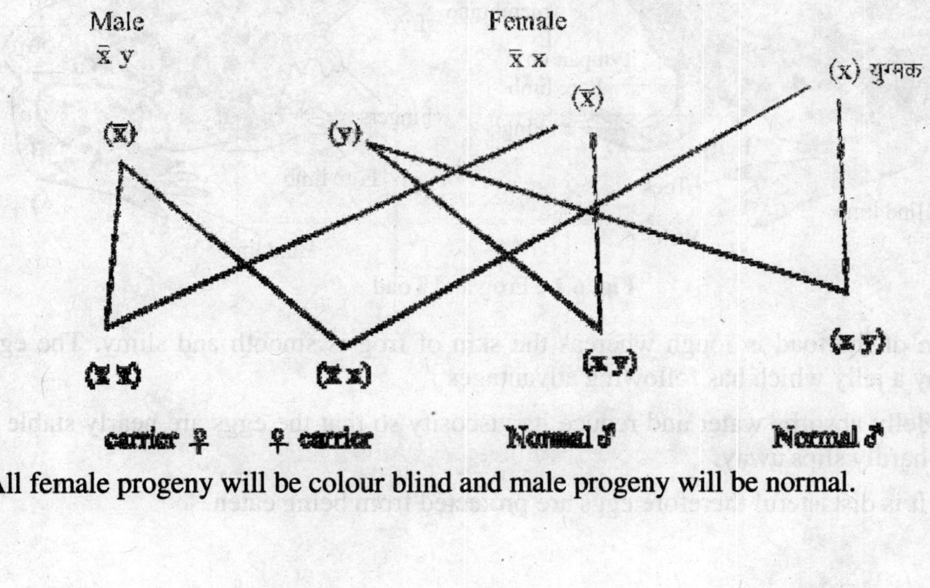

All female progeny will be colour blind and male progeny will be normal.

Unit - 6
Developmental Biology

1. STUDY OF DEVELOPMENT OF FROG/TOAD

Frog and Toad comes in class *Amphibia* and order *Salientia.* Frogs are classified in the family *Ranidae* and toads in *Bufonidae.* Life cycle of both species can be easily studied in rainy season. Author himself reared and collected the developmental stages of toad from a temporary pond in the college campus. The frog we generally see in the ponds and other water bodies is *Rana tigerina*, *Rana cynophylctis* is also commonly found. *Rana tigerina* can be easily differentiated from *Bufo melanostictus*. This differentiation is prominent in eggs and tadpole stages. The tadpoles of toads are comparatively smaller than the tadpoles of frog. The eggs of toad are in the form of a chain.

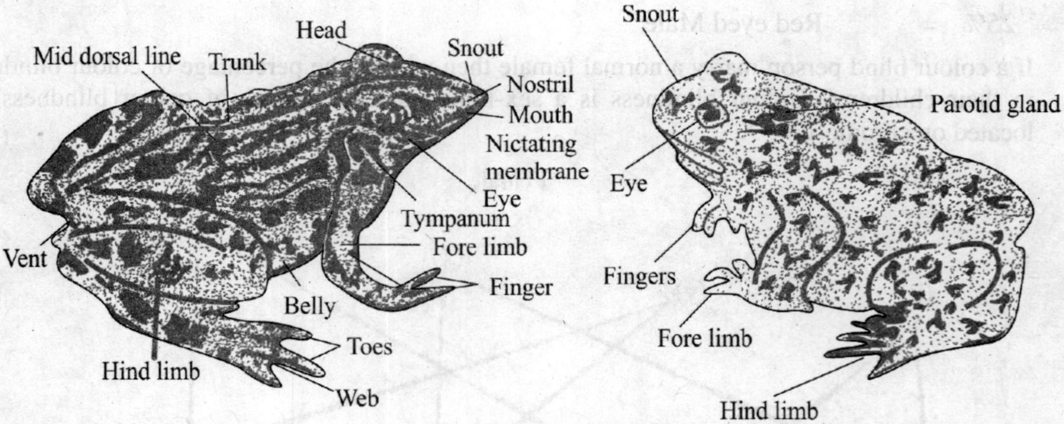

Fig. 6.1 : Frog and Toad

The skin of the toad is rough whereas the skin of frog is smooth and slimy. The eggs are surrounded by a jelly which has following advantages :

1. Jelly absorbs water and reduce its viscosity so that the eggs are nearly stable and is hardly slips away.
2. It is distasteful therefore eggs are protected from being eaten.

3. It is transparent and the sun rays easily penetrate through it and provide necessary energy.

4. It protects the eggs from being blown as it easily attach to the objects or vegetation in water.

The eggs are *telolecithal* (The yolk is concentrated on one pole called vegetal pole, the pigmented part is animal pole).

Various stages of Development of Frog/Toad

1. Egg

The eggs of frog are in the form of bunch or spawn whereas they are in beaded chain or streak in toads. They can be easily collected in rainy season and preserved in 2% formalin or 70% alcohol.

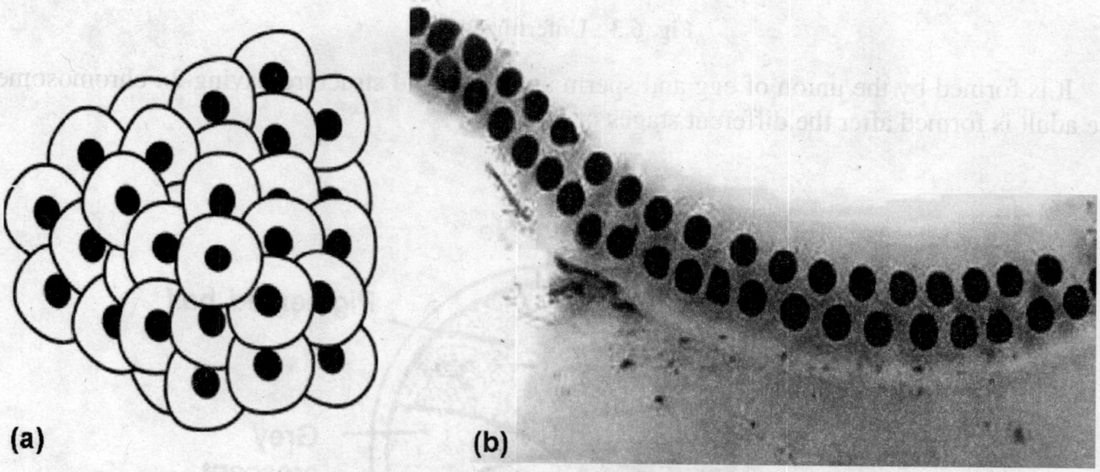

Fig. 6.2 : Spawn of toad (b) and frog (a)

The unfertilized egg is a spherical structure of 1.6 mm diameter. A vitelline membrane covered by albumin is present. Egg nucleus is located in animal pole. The nutrition during development is provided by the yolk.

Fertilized egg

It is similar to unfertilized egg but grey crescent is found. Fertilized egg appears to be divided into three parts -

(a) upper pigmented part

(b) grey crescent part (light black)

(c) lower yolky part.

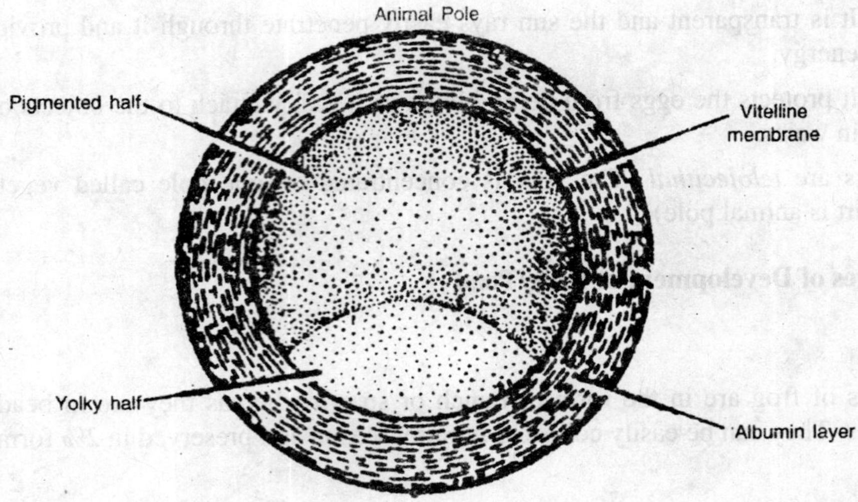

Fig. 6.3 : Unfertilized egg

It is formed by the union of egg and sperm and is diploid structure having 2x chromosomes. The adult is formed after the different stages of cleavage.

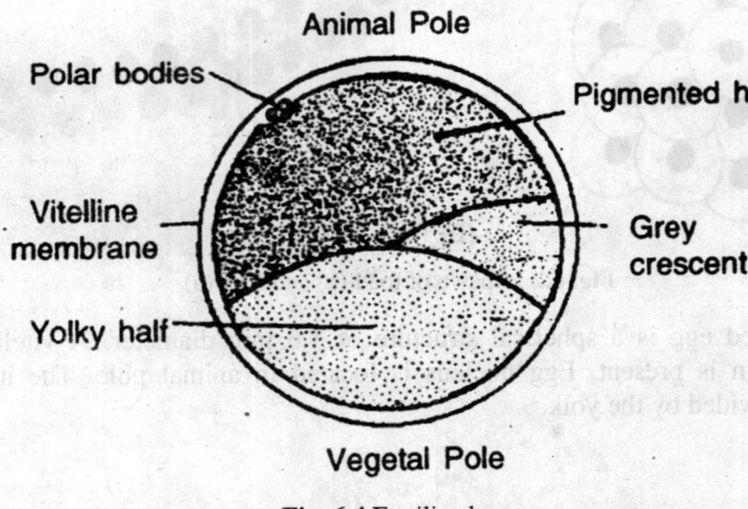

Fig. 6.4 Fertilized egg

2. Cleavage

Cleavage starts 2-3 hrs. after fertilization and is holoblastic and complete, the first cleavage is vertical which divides the egg into two cells, thereafter the eggs divides into 4, 8, 16, 32, 64 and 128 cells. The cleavage then become irregular. The eggs look like a mulberry. This stage is *morula*.

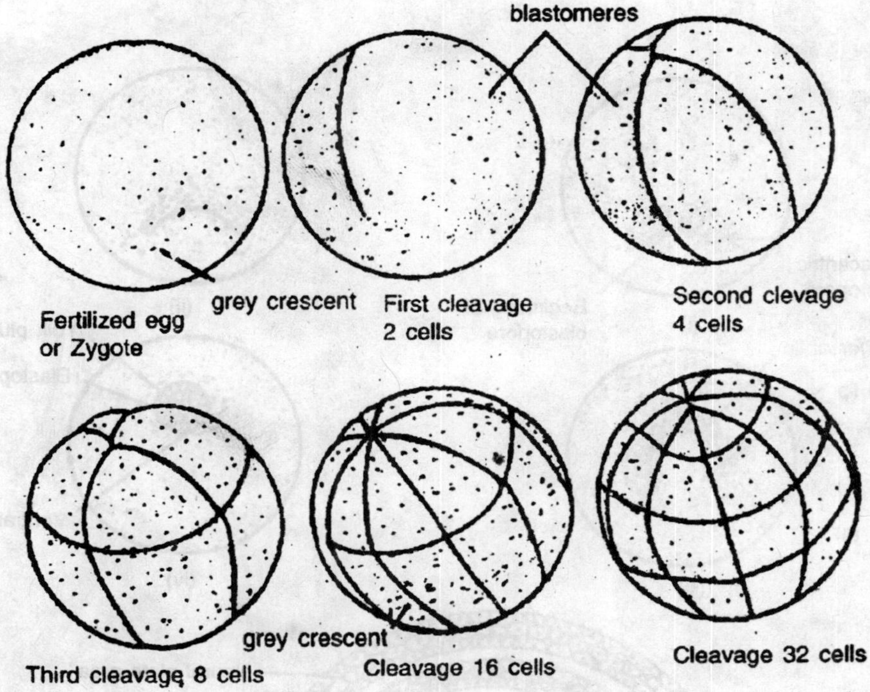

Fig. 6.5: Stages in cleavage

3. Blastula

Blastocoel start its formation at the 8 celled stage of cleavage which becomes distinct after morula stage. A liquid is filled in it and look like a hollow ball. Its base is made up of large yolk cells called *macromeres* and upper smaller pigmented cells are called *micromeres*. Blastocoel is little bit eccentric at the animal pole.

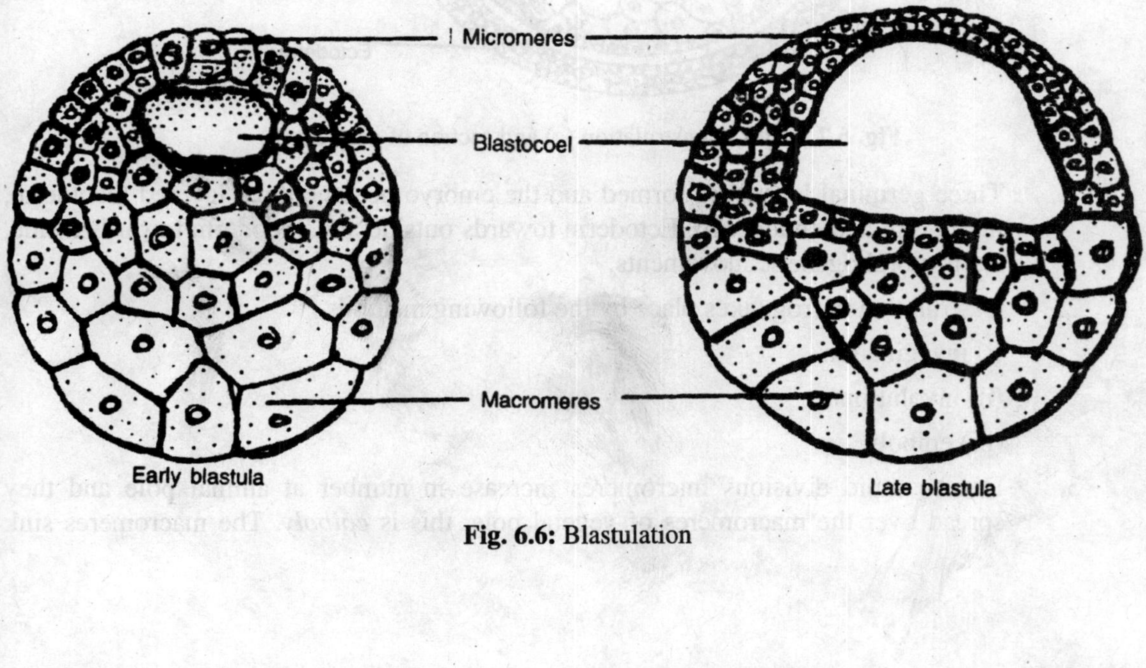

Fig. 6.6: Blastulation

4. Gastrula

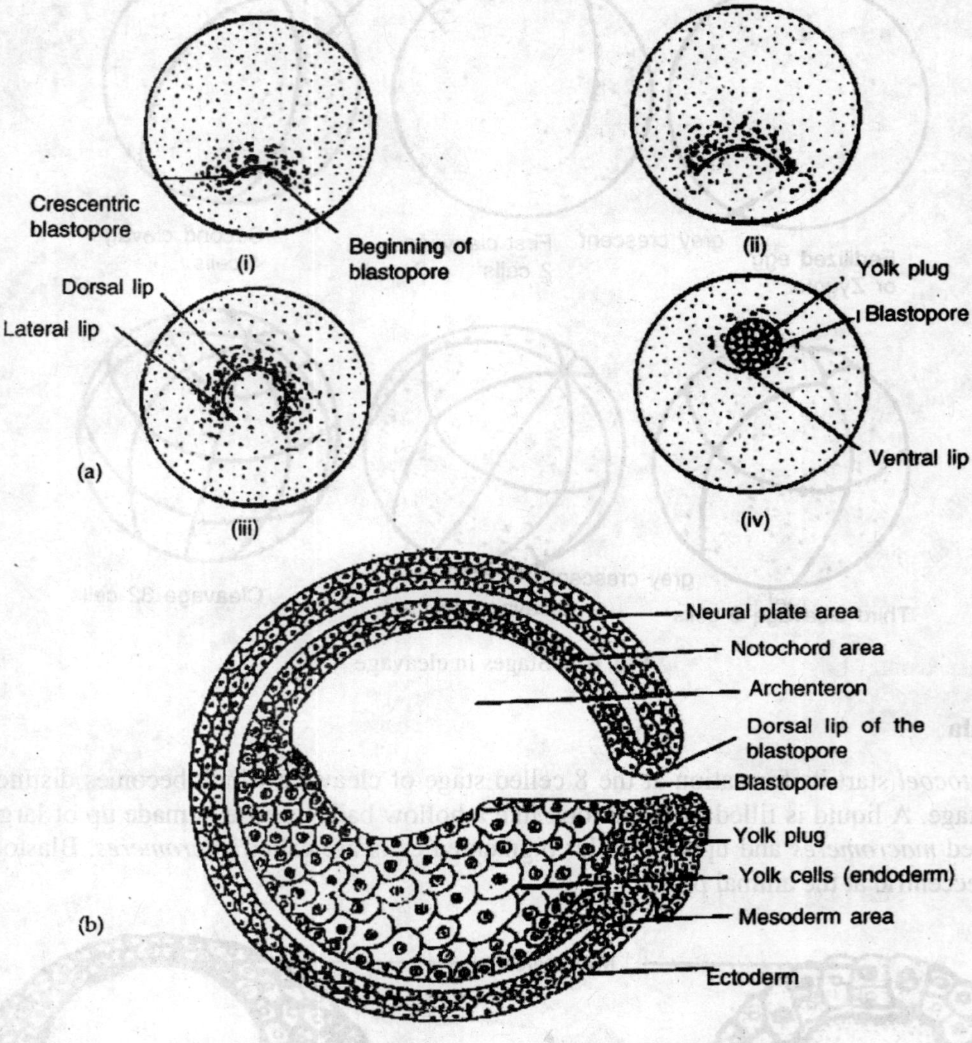

Fig. 6.7 Stages in gastrulation (a) and section of gastrula (b)

1. Three germinal layers are formed and the embryo become three layered (ecotderm, mesoderm and endoderm). Ectoderm towards outside and endoderm towards inside, starts morphogenetic movements.
2. Gastrulation in frog takes place by the following methods :

   (i) invagination

   (ii) involution

   (iii) epiboly
3. Due to rapid divisions micromeres increase in number at animal pole and they spread over the macromeres of vegetal pole, this is *epiboly*. The macromeres sink

towards the blastocoel it is called *invagination. Gastrocoel,* is formed by invagination. It froms alimentary canal and liver.

4. A pore called *blastopore* is formed at the invagination. Its upper end is called *dorsal lip* which on growing becomes crescentric, the dorsal lip grown in lateral side and become horse shoe shaped, it is called *lateral* lip. Macromeres covers the blastopore called *yolk plug* (Fig. 6.7 a iii, iv).
5. The cells from dorsal lip grow towards interior of blastocoel, it is *involution*.
6. Involution reduces the area of blastocoel and archenteron increases in size. Blastocoel ultimately disappear.

Vertical Section of Gastrula

(a) Early Gastrula

1. It appears like a ball.
2. Micromeres cover the macromeres.
3. Invagination dorsolaterally indicate the beginning of blastopore.
4. Blastocoel is covered by micromeres and macromeres dorso-ventrally.

(b) Late Gastrula

1. Micromeres and macromeres are conspicuous.
2. Micromeres due to involution, reduce blastocoel and the archenteron appears.
3. Blastopore appears dorsolaterally because of invagination.
4. Macromere forms the base of archenteron.
5. Embryo becomes three layered (Fig. 6.7).

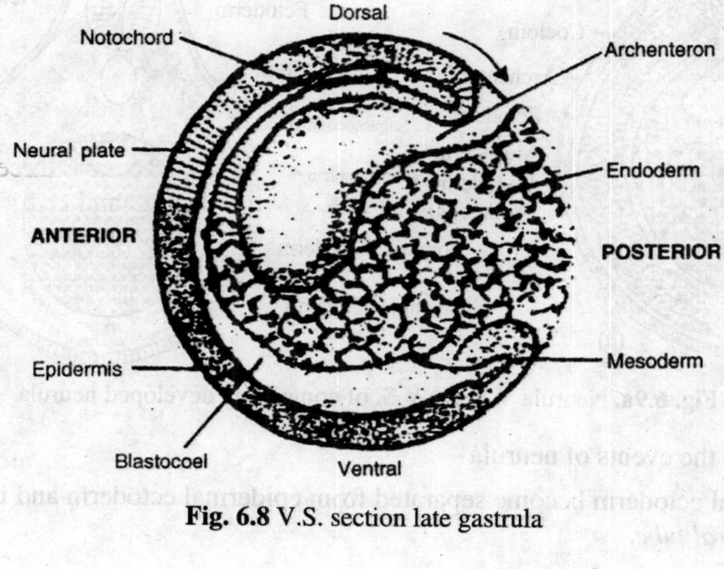

Fig. 6.8 V.S. section late gastrula

5. Neurula

Formation of neural tube (which form brain and spinal cord) is called *neurulation* and the embryonic stage is called neurula.

Neuroectoderm becomes pear shaped mid dorsally. It forms neural plate. Neural plate grows from the margins to form a neural fold. A neural groove appears between the fold which become deep by increasing the neural folds. When both the folds unite each other a *neural tube* is formed (Fig. 6.9).

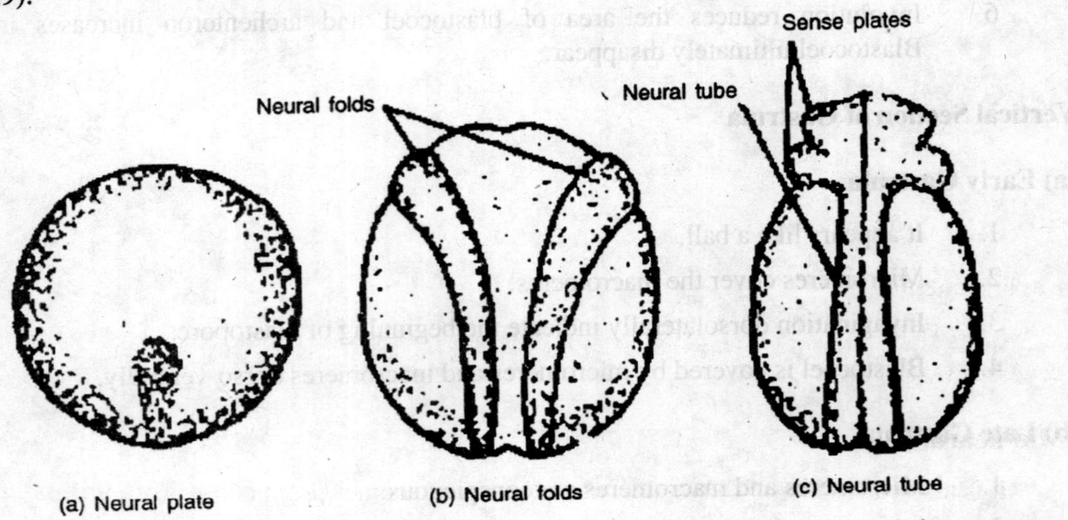

Fig. 6.9 Different stages of neurulation

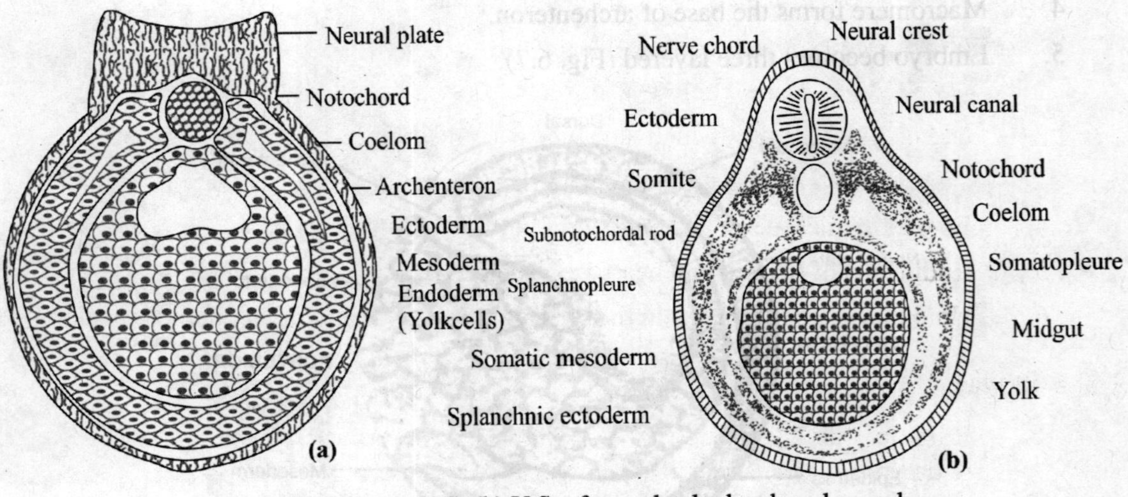

Fig. 6.9a. Neurula V.S. (b) V.S. of completely developed neurula

Following are the events of neurula -

1. Neural ectoderm become separated from epidermal ectoderm and invaginate to form a *neural tube.*

2. Mesoderm and chordamesoderm become separated from endoderm.
3. Mesoderm is differentiated into three parts.
   (a) Dorsal mesoderm/Epimere
   (b) Mesomere
   (c) Hypomere/latero-ventral mesoderm

Epimere give rise to dermis, sclerotomes and myotomes (sclerotome and myotomes give rise to vertebral column and skeletal muscles).

Mesomere give rise to reproductive and excretory organs and *Hypomere* to coelom, heart, blood cells and blood vessels.

The neurula develops through

– early neurula
– mid neurula
– late neurula

Early Neurula

It is also called neural plate stage. The neural plate is nearly flat and the neural folds develops (Fig. 6.9a).

Mid neurula

It is also called neural fold stage. Neural groove becomes prominent (Fig. 6.9b).

Late Neurula

Neural tube develops and becomes conspicuous (Fig. 6.9c).

VLS of Neurula (Vertical longitudinal section)

1. Notochord is located below neural tube.
2. Mesodermal somites which develops from mesoderm, appears.
3. The anterior end of neural tube begin to differentiate into brain.
4. Ectoderm, mesoderm and endoderm becomes prominent.
5. Rudiments of thyroid and heart are visible.
6. Upto this stage it remains in the jelly therefore anus and mouth do not develop.

Tail Bud Stage

1. Prior to 4th day of fertilization, the embryo grows posteriorly and a bud like structure appears. It is called *tail bud* and the stage is tail bud stage.
2. It grows continuously to form the tail of the tadpole.
3. Ventral and dorsal fin develops on the ventral and dorsal margins of the tail respectively. Myotomes and notochord are also develops later on.

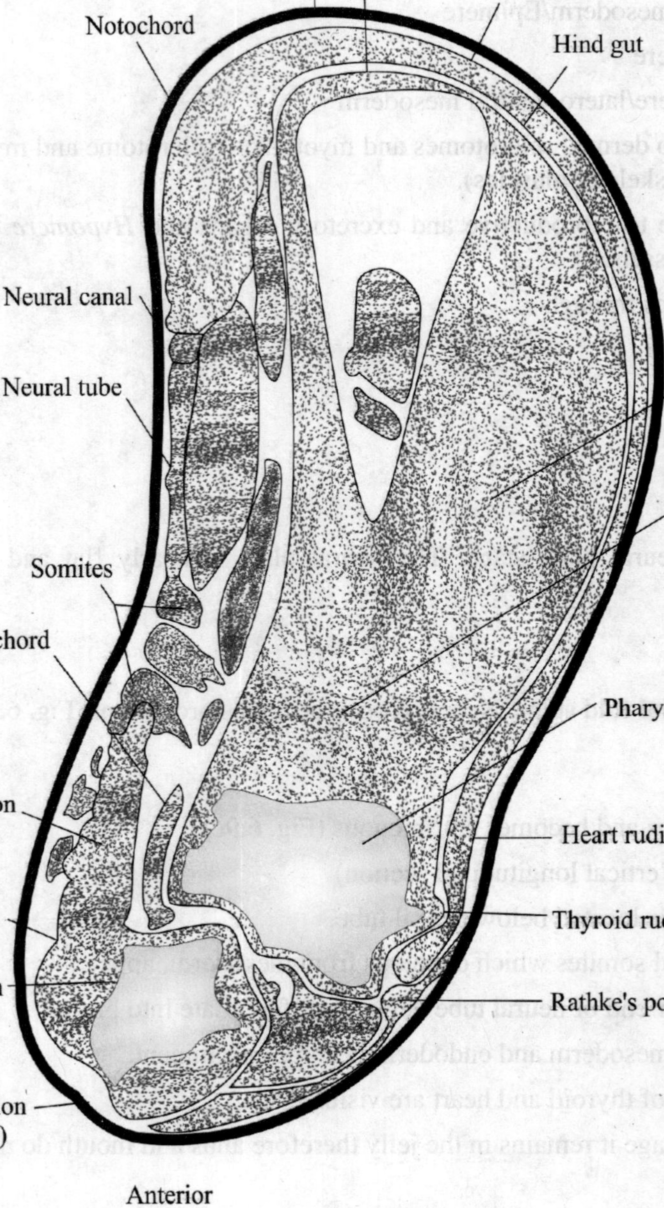

Fig. 6.10 V.L.S. of neurula

4. Head bears a mucus gland which functions as oral sucker by which the tadpole anchors to aquatic vegetations.
5. Optic bulge with middle of the head appears.

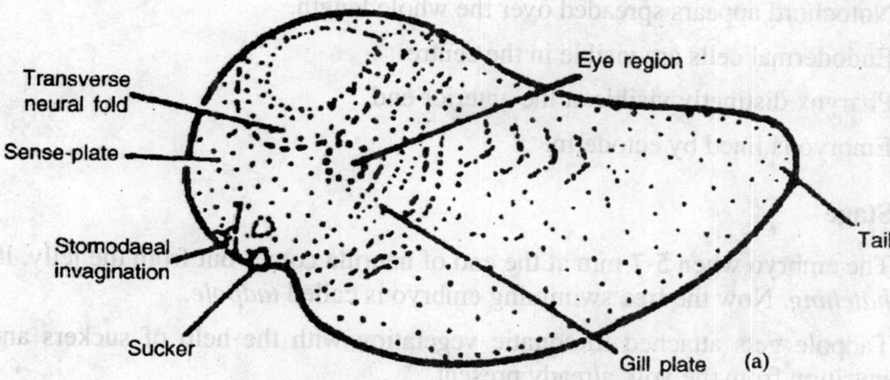

Fig. 6.11a. Frog embryo : tail bud stage

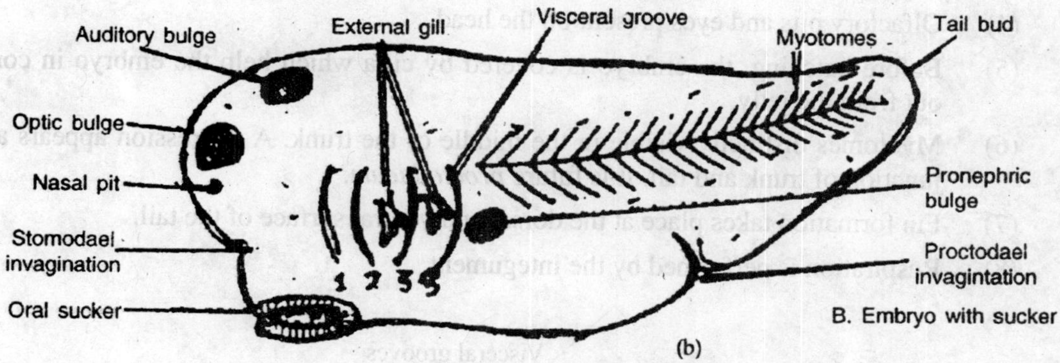

Fig. 6.11b. Tail bud stage : internal structure

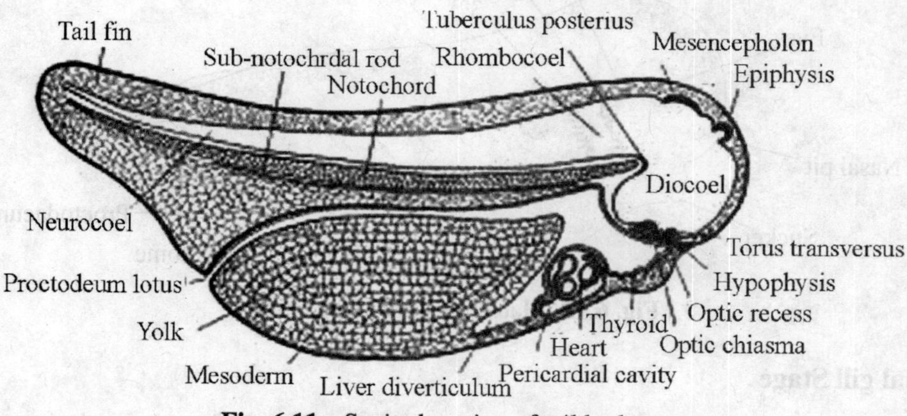

Fig. 6.11c. Sagittal section of tail bud stage

Section of Tail bud stage

(1) Neural tube at the anterior end, is differentiated into prosencephalon, mesencephalon and rhombencephalon.

(2) Liver diverticula are formed by the cells of archenteron which later on differentiated into liver.

(3) Notochord appears spreaded over the whole length.

(4) Endodermal cells are visible in the centre.

(5) Pharynx distinctly visible at the anterior end.

(6) Embryo is lined by ectoderm.

7. Hatching Stage

(1) The embryo when 5-7 mm at the end of neurula comes out from the jelly. It is called *hatching*. Now the free swimming embryo is called *tadpole*.

(2) Tadpole gets attached to aquatic vegetation with the help of suckers and derives nutrition from the yolk already present.

(3) Mouth not present at the time of hatching.

(4) Olfactory pits and eyes present on the head.

(5) Before hatching, the embryo is covered by cilia which help the embryo in coming out from the jelly.

(6) Myotomes distinctly visible in the middle of the trunk. A depression appears at the junction of trunk and tail. It is future *proctodaeum*.

(7) Fin formation takes place at the dorsal and ventral surface of the tail.

(8) Respiration is performed by the integument.

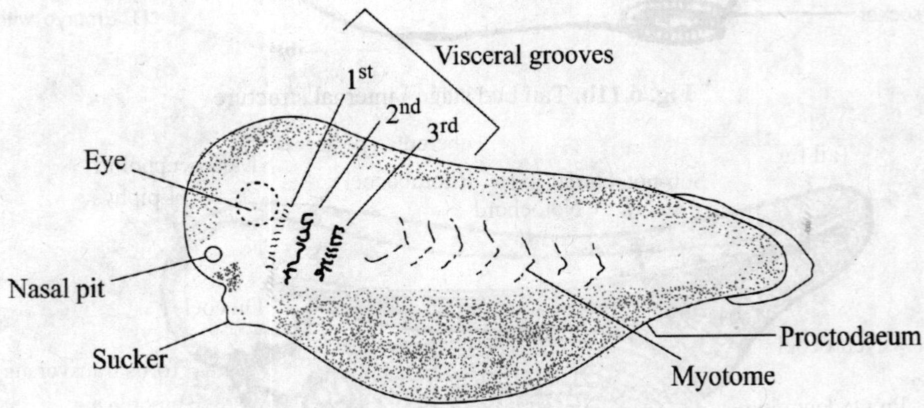

Fig. 6.12 : Hatching stage : frog embryo

8. External gill Stage

(1) Though the formation of external gills takes place before the embryo escapes from the jelly but they are distinctly visible in the tadpole attached with aquatic vegetation.

(2) Two pair of external gills on both side of the head are conspicuous, the third pair is very small and appears attached with the 2nd pair.

(3) External gills are formed by the IIIrd, IVth and Vth visceral arch

(4) The blood is brought by IIIrd, IVth and Vth afferent branch of aortic arch and carreid away by efferents.

(5) The exchange of O_2 and CO_2 is accomplished by the circulating blood in the gill filaments. The oxygen is taken up and CO_2 is given out by diffusion.

(6) *Pronephric* kidneys developed completely and become functional.

(7) *Stomodaeum* on the sucker develops at the anterior end and its opening becomes the mouth.

(8) Horny jaws bearing teeth are found around the mouth.

(9) Alimentary canal becomes coiled like a spring.

(10) Olfactory pits and lens in the eyes are prominent.

(11) *Opercular fold* begins to develop.

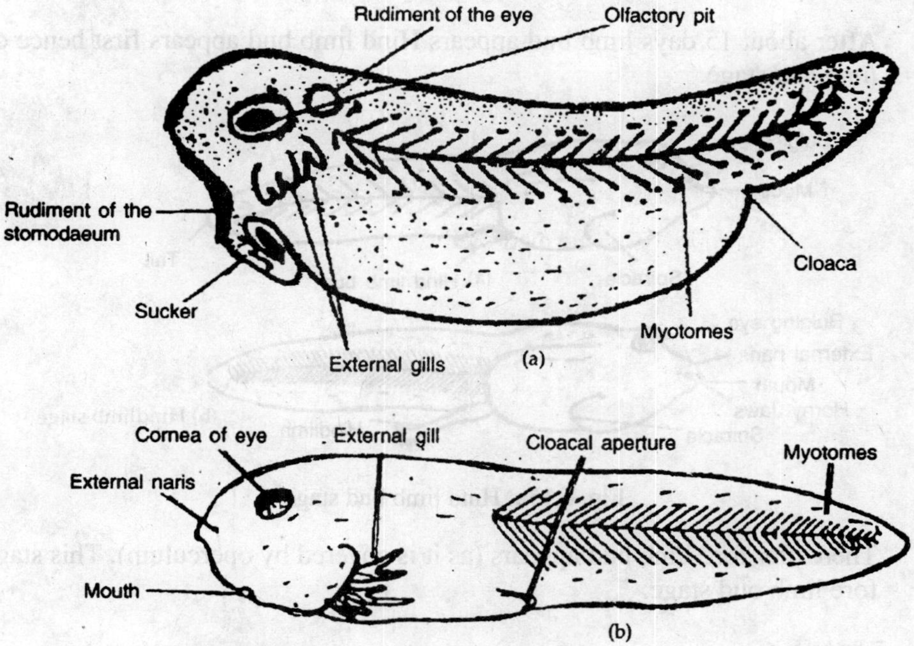

Fig. 6.13a & b. External gill stage

9. Internal gill Stage

(1) The external gill stage is not long lasting and internal gills starts to develop very soon. Hind branch of IIIrd, IVth and Vth visceral arch projects out into finger like projections called internal gills. The tadpole now respire through internal gills.

(2) Water enters from mouth and circulates through opercular chamber. It is out from the spiracle.

(3) Sucker and adhesive glands disappear.

(4) The tail becomes laterally compressed. Ventral, dorsal and caudal fins develops.

(5) Rows of denticles found in the mouth.

(6) *Hind limb bud* appears at the junction of trunk and tail.

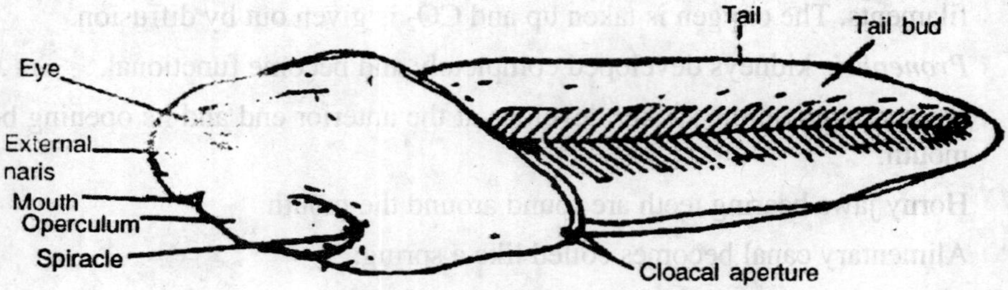

Fig. 6.14. Internal gill stage

10. Limb bud Stage

(1) After about 15 days limb bud appears Hind limb bud appears first hence called *hind limb bud* stage.

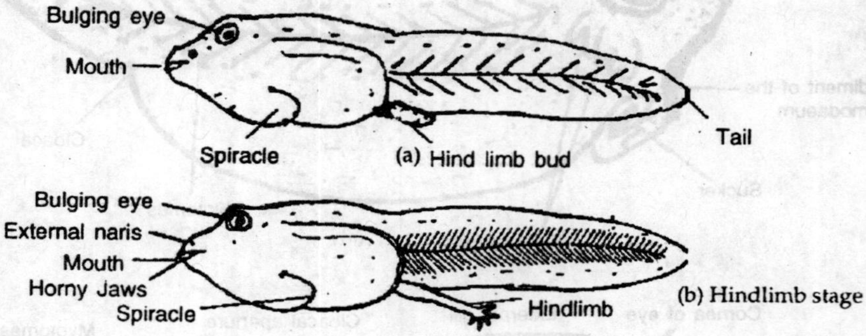

Fig. 6.15a. Hind limb bud stage

(2) Thereafter fore limb bud appears (as it is covered by operculum). This stage is called fore limb bud stage.

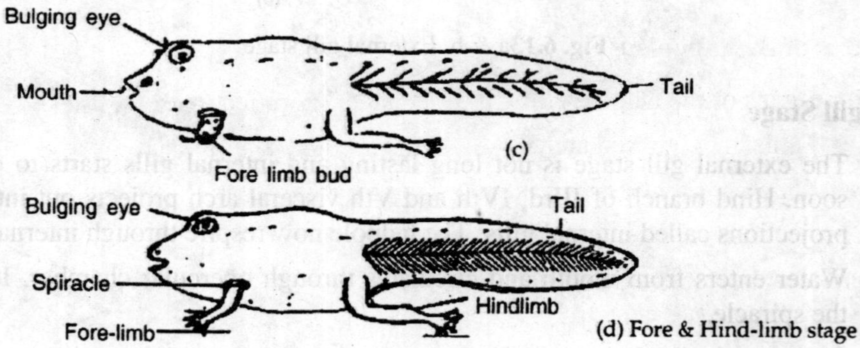

Fig. 6.15c (Fore limb bud stage)

Mature tadpole Stage

(1) Hind limb bud which is stump like gets differentiated into fingers.

(2) Both hind limbs grow and becomes enlarged.

(3) Thereafter forelimb appears (Left limb appears first then right limb because the spiracle on left remains open and the limb through it appears first).

(4) Mature tadpole is herbivorous and the respiration is accomplished by internal gills.

(5) Water enters into the mouth and comes out from the spiracle through pharynx.

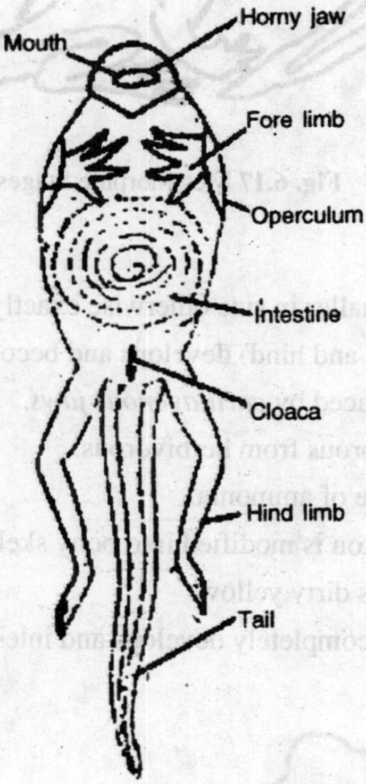

Fig. 6.16. Mature tadpole

12. Metamorphosis of the tadpole

(1) Major organs of the tadpole either degenerate or modified into new ones.

(2) Internal gills are absorbed and the gill cleft are closed.

(3) Mouth which is circular now becomes laterally elongated.

(4) Tongue enlarges in size.

(5) Legs becomes elongated and stout.

(6) The eyes appears protruded on the dorsal surface of the head.

(7) Eye lids and nictating membrane are formed.

(8) Long and coiled intestine becomes short and straight.

(9) Tail is degenerated slowly and totally disappears by the end of metamorphosis.

(10) Aquatic tadpole now becomes terrestrial.

Fig. 6.17 Metamorphic Stages

Froglet/Toadlet

(1) Froglet/toadlet is smaller in size otherwise exactly similar to the adult.

(2) Both the limbs (fore and hind) develops and becomes stout.

(3) Horny jaws are replaced by *cartilagenous jaws.*

(4) It becomes insectivorous from herbivorous.

(5) Excrete urea in place of ammonia.

(6) Cartilagenous skeleton is modified into bony skeleton.

(7) Integument becomes dirty yellow.

(8) Liver and pancreas completely develops and intestine becomes short.

Fig. 6.18 : Froglet

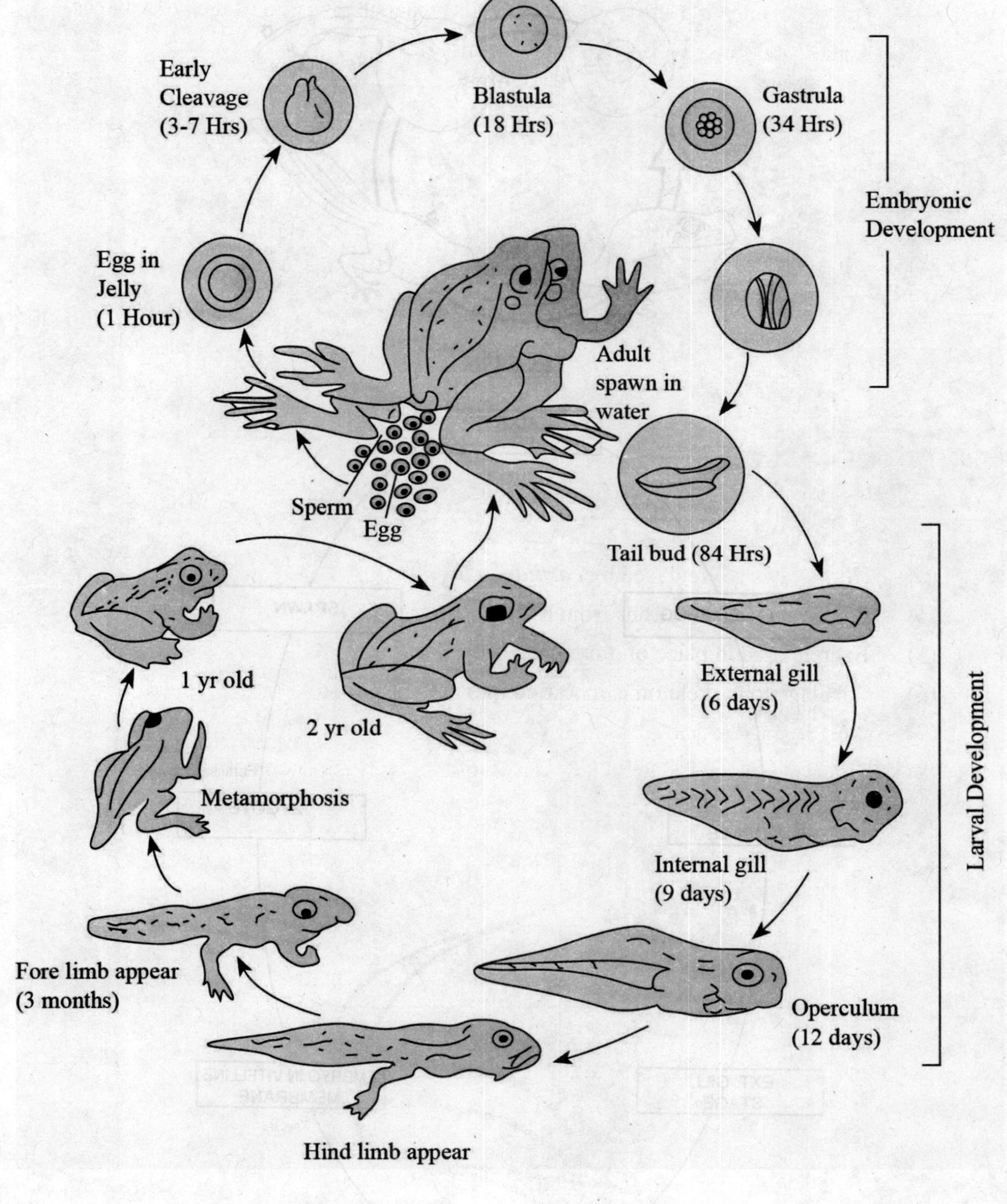

Fig. 6.19a : Life cycle of frog

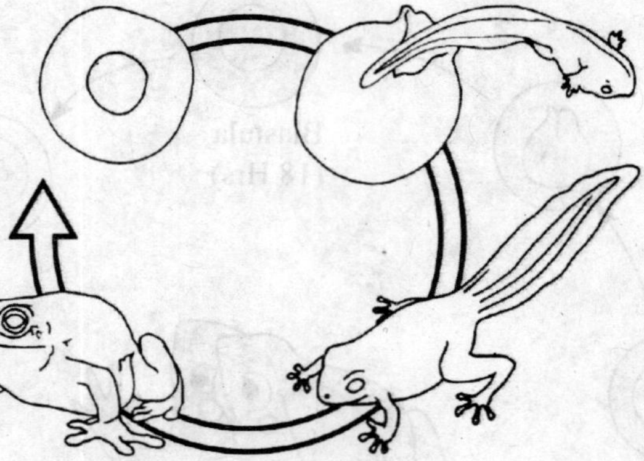

Fig. 6.19b.

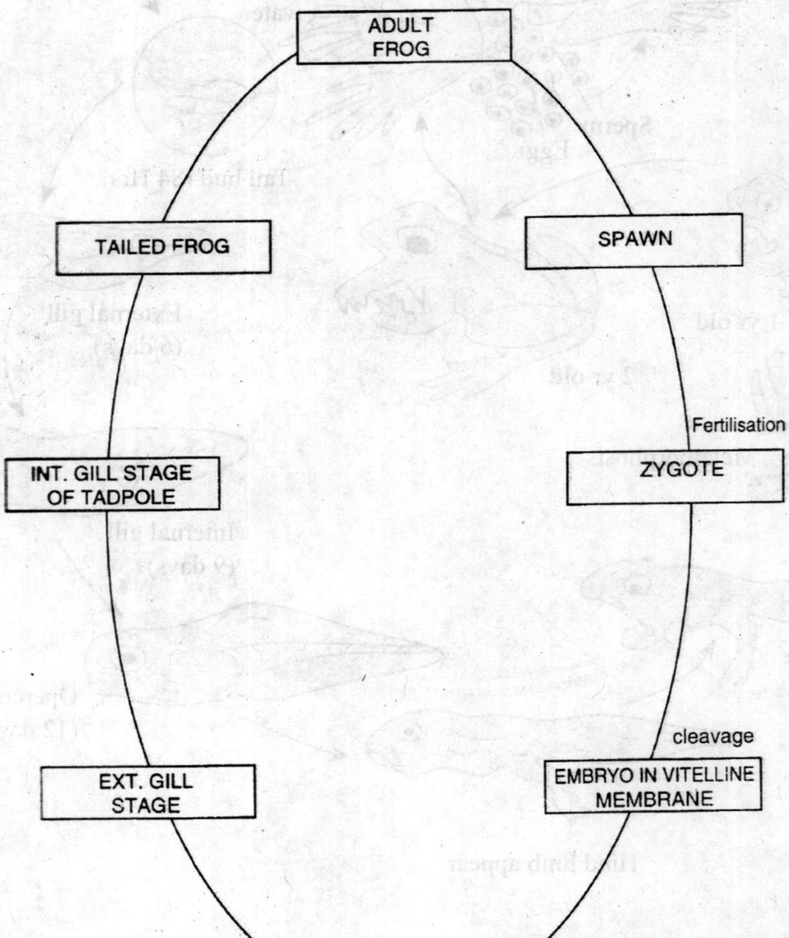

Fig. 6.20 : Graphic representation of life cycle

2. Various Stages of Development of Chick

1. Structure of hen's Egg

(1) Hen's egg is megalecithal (yolk in enormous amount).

(2) It is the best example of *cleidoic egg* (self sufficient).

(3) External shell is made up of Calcium carbonate ($CaCO_3$)

(4) Underlying shell, are two membranes which are separated at broader end by the air space.

(5) Vitelline membrane surrounding the yolk present which is covered by albumin.

(6) When the egg descends down into the oviduct, the albumin becomes cord like called *chalaza.*

(7) White yolk is found in the middle of the egg called *latebra.*

(8) Blastoderm is found over the *latebra.*

(9) Concentric rings of yellow and white yolk are present around the *latebra.*

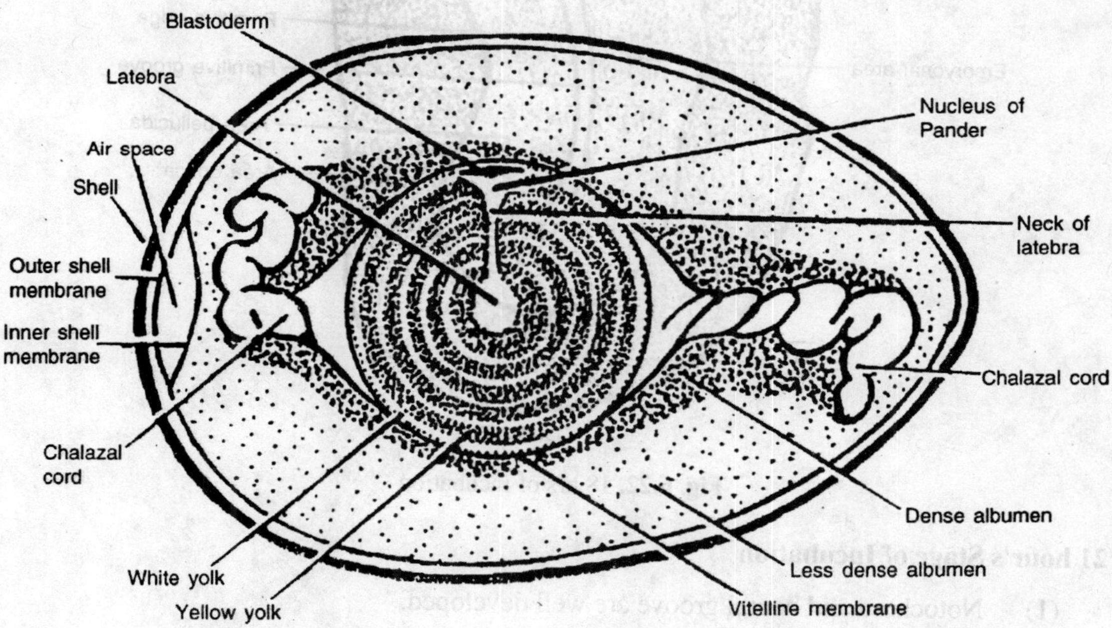

Fig. 6.21. Hen's egg

2. 18 hour's Stage of Incubation

(1) After 18 hours of incubation, *Primitive streak,* notochord and neural plate becomes conspicuous.

(2) Pro-amnion develops on the anterior end of neural plate.

(3) Neural plate is separated from primitive streak by Hensen's node.

(4) Area opaca and area pellucida are prominent.

(5) It is also called head process stage.

Special Remark

After the formation of primitive streak, the cells near the Hensen's node starts forming notochord.

The anterior part of neural plate in the cephalic region is called pro-ammion.

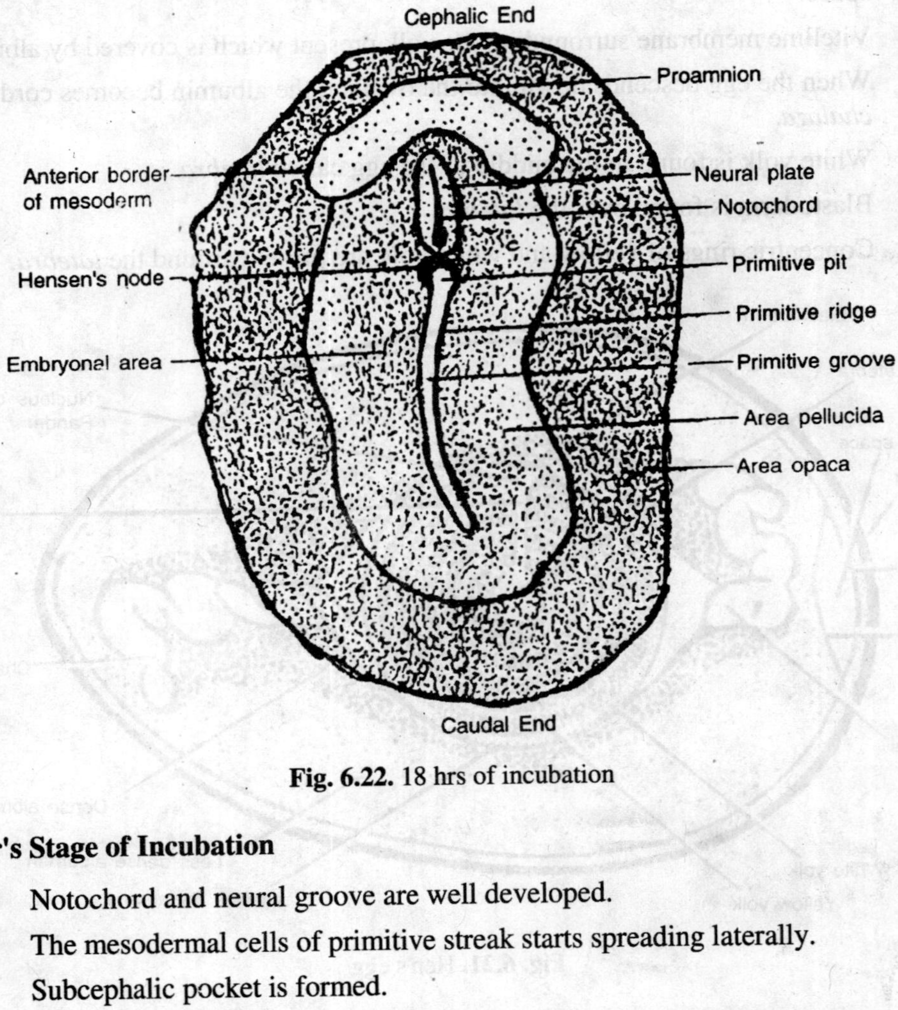

Fig. 6.22. 18 hrs of incubation

3. 21 hour's Stage of Incubation

(1) Notochord and neural groove are well developed.

(2) The mesodermal cells of primitive streak starts spreading laterally.

(3) Subcephalic pocket is formed.

(4) Neural groove is separated from primitive streak by Hensen's node.

(5) Area pellucida forms embryonic and extra embryonic area

(6) One pair of mesodermal somites appear. Thereafter one pair of somite develops for every hour.

(7) Formation of fore-gut starts from the differentiating cells of primitive gut.

Special Remark

From the 20 hours incubation stage, the formation of foregut starts from the differentiating cells of primitive gut.

Mesoderm of dorsal surface starts thickening which forms mesodermal somites later on.

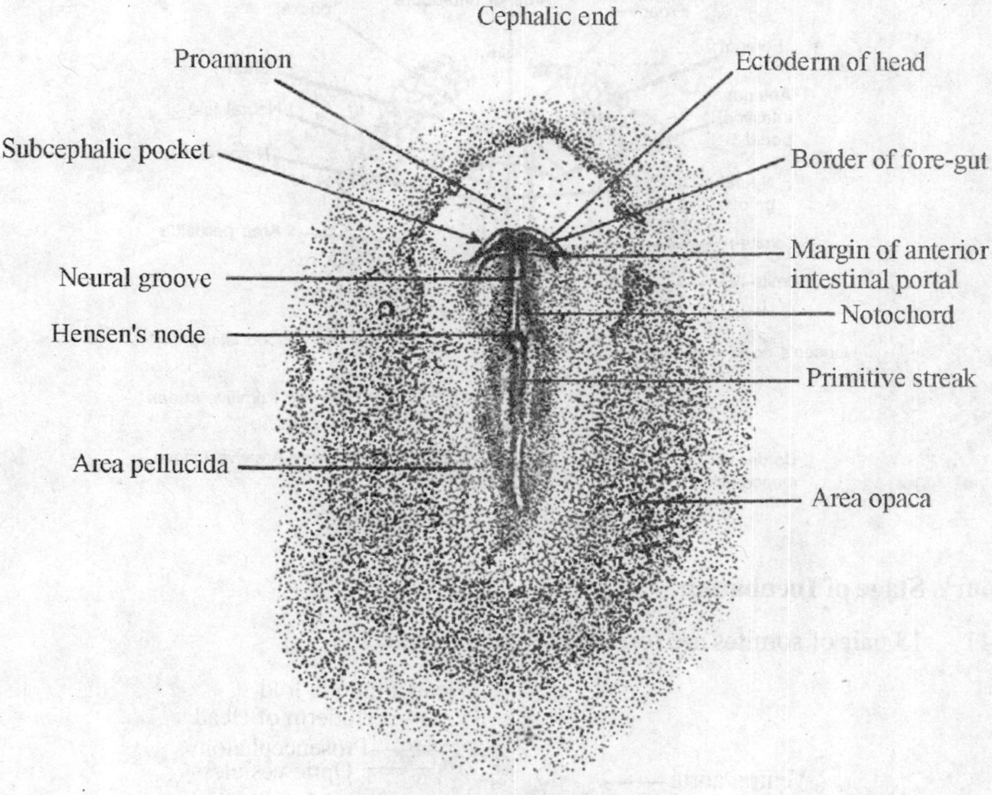

Fig. 6.23. 21 hrs of incubation

. 24 hour's Stage of Incubation

(1) Embryonic and extra embryonic regions are conspicuous.

(2) Indistict primitive streak is displaced towards caudal end.

(3) Area vasculosa develops in area opaca. Mesodermal cells divided into smaller groups of blood islands.

(4) The area of opacal region without capillaries is called area vitellina.

(5) Head fold appears on the anterior end which is formed by the lifting of ectoderm.

(6) Endoderm becomes coiled to form fore gut in anterior part.

(7) Notochord appears rod like.

(8) 4 pair of mesodermal somites present (3 pairs distinct and one indistinct).

Special Remark

The anterior part of the embryo develops rapidly from 22 hours of incubation therefore it appears little bit lifted.

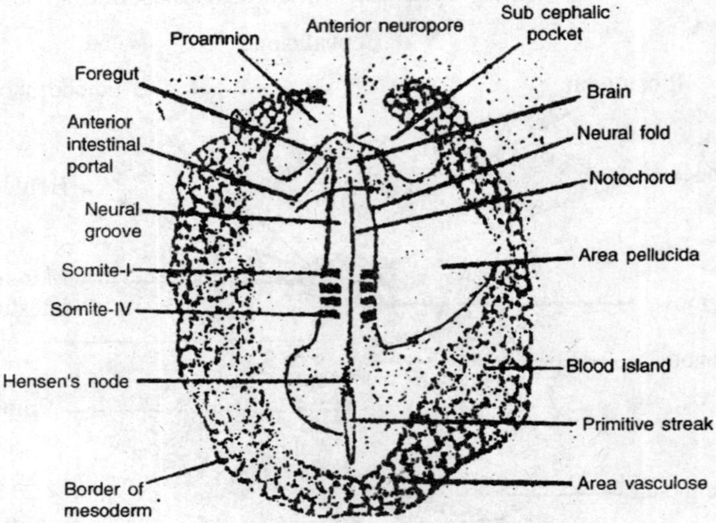

Fig. 6.24. 24 hrs of incubation

5. 33 hour's Stage of Incubation

(1) 13 pair of somites are formed (12 pairs distinct)

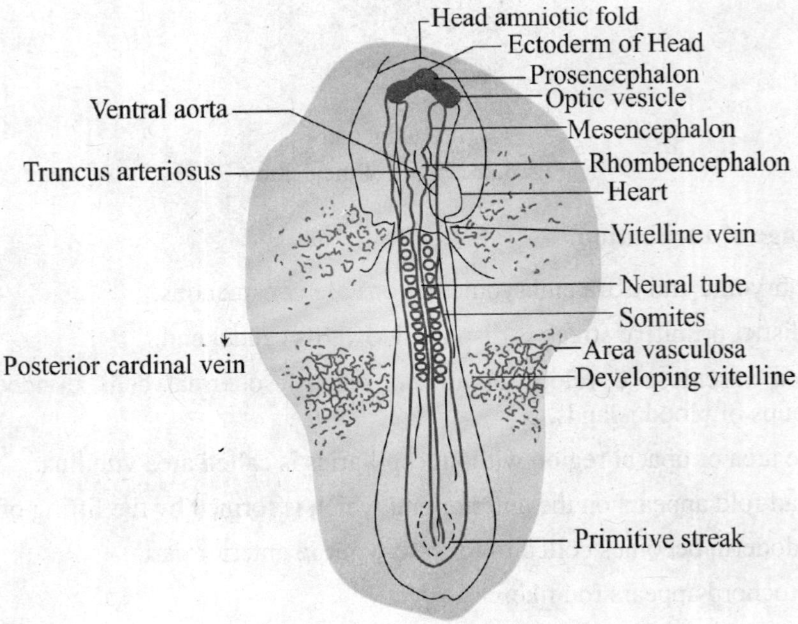

Fig. 6.25. 33 hrs of incubation

(2) The brain is divided into fore, mid and hind brain.

(3) Heart starts developing.

(4) Primitive streak becomes small.

(5) Optic vesicles are visible as lateral outgrowths in the anterior brain.

(6) Brain, notochord is clearly visible between somites and neural tube.

(7) Omphalo mesentric vein appears in the anterior region.

Special Remark

The development of circulatory and central nervous system takes place from this stage.

6. 48 hour's Stage of Incubation

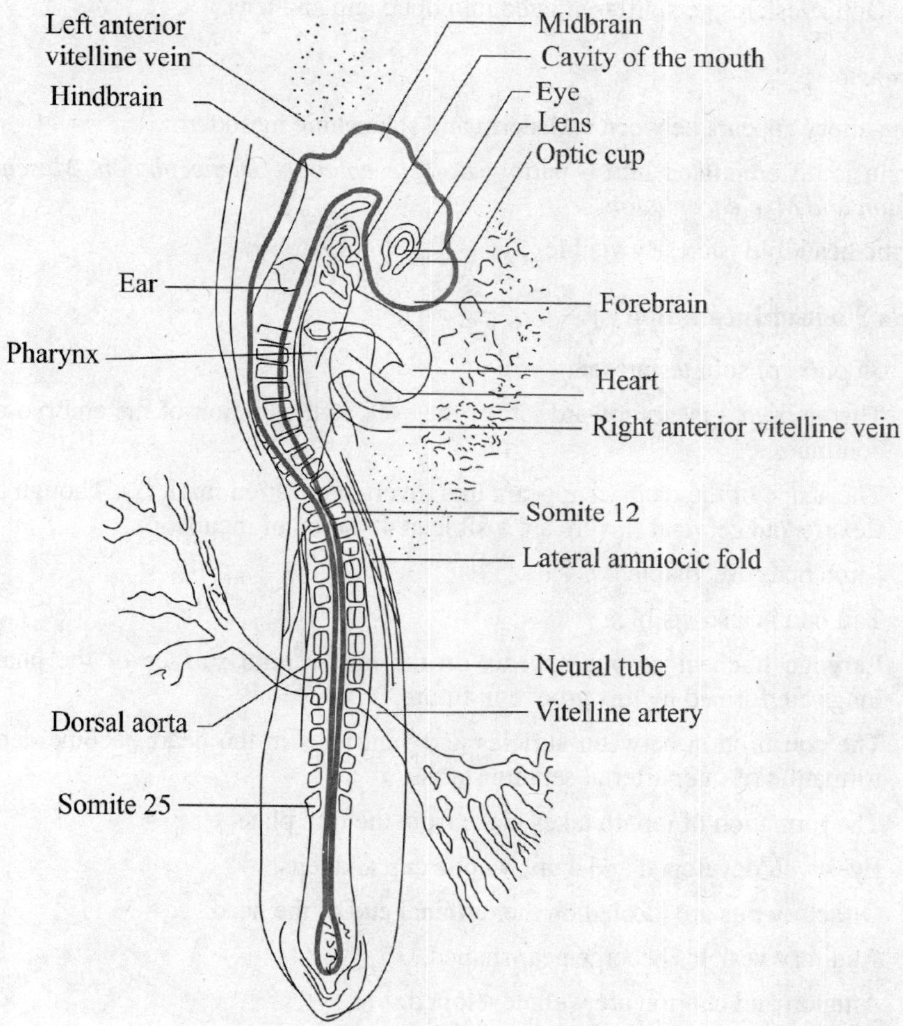

Fig. 6.26. 48 hrs of incubation

(1) *28 pairs* of somites present.

(2) The head appears tilted to right. The changes in the gross anatomy of embryo due to torsion and flexure, are clearly visible.

(3) The fore and hind brain appears close as the neural tube in the anterior part turn in mesencephalon region.

(4) Hind brain also exhibit similar bending. It is called *cervical flexure*. The embryo appears `*f*' shaped.

(5) Vascular net increases in the pellucida, heart becomes differentiated into auricles, ventricles, sinus venosus and truncus arteriosus.

(6) Primitive streak becomes small and indistinct.

(7) Auditory vesicles or otocyst are visible.

(8) Optic vesicles gets differentiated into optic cup and lens.

Special Remark

Vitelline artery appears between endoderm and splanchnic mesoderm.

The brain is differentiated into 5 parts viz., *Telecephalon, Diencephalon, Mesencephalon, Metencephalon and Myelencephalon.*

Ammiotic head fold is clearly visible.

7. 72 hour's Stage of Incubation

(1) *35 pairs* of somites present.

(2) The embryo appears toward left of the yolk as the torsion of the embryo on its axis continues.

(3) The shape of the embryo appears like inverted question mark (?). Though the cranial flexure and cervical flexure are visible at 48 hours of incubation.

(4) Limb buds are visible.

(5) Tail bud is also visible.

(6) Laryngo tracheal groove appears on the mid ventral surface of the pharynx. The lungs are formed by this groove in future.

(7) The constriction between auricles and ventricles in the heart become deep and the formation of inter arterial septum starts.

(8) The formation of mouth takes place from the oral plate.

(9) Eyes well developed and exhibit optic cup and lens.

(10) Olfactory pits are located on the terminal end of the head.

(11) Auditory vesicles become pear shaped.

(12) Amnion and chorion are well developed.

(13) Allantois is formed by the outgrowth of the hind gut of embryo.

Special Remark

4th visceral arc appears from the 3rd day of incubation.

Torsion of the body completed except tail region.

The central part of the embryo on dorsal surface becomes concave.

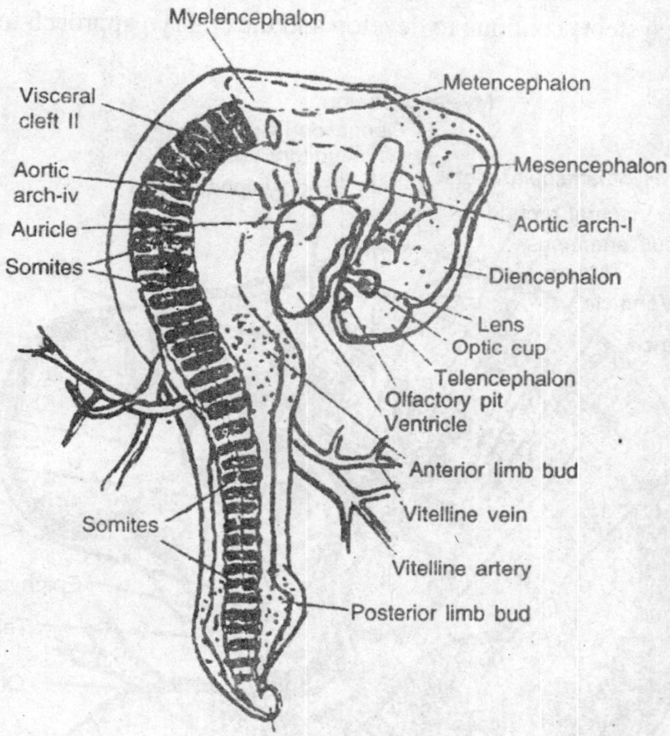

Fig. 6.27. 72 hrs of incubation

8. 96 hour's Stage of Incubation

(1) *41 pairs* of somites are present.

(2) The entire body due to torsion, rotates through 90° and the embryo lie on the left side of the yolk.

(3) Body clearly divided into head, trunk, and tail.

(4) Due to cranial, lumber and caudal flexure, the embryo bends to its left and become `C' shaped, as a result head and tail comes in the proximity.

(5) Different regions of the brain are clearly visible.

(6) The sensory organs (olfactory, optic and auditory) begins to develop.

(7) First and second aortic arch become degenerated and 5th and 6th arch are completely developed.

(8) Different parts of the heart develops and enlarged.

(9) Amnion and chorion are completely developed. Allantois and yolk sac are surrounded by chorion.

Special Remark

Cranial and spinal nerves develops.

Different organ systems continue to develop and the embryo approach towards perfection.

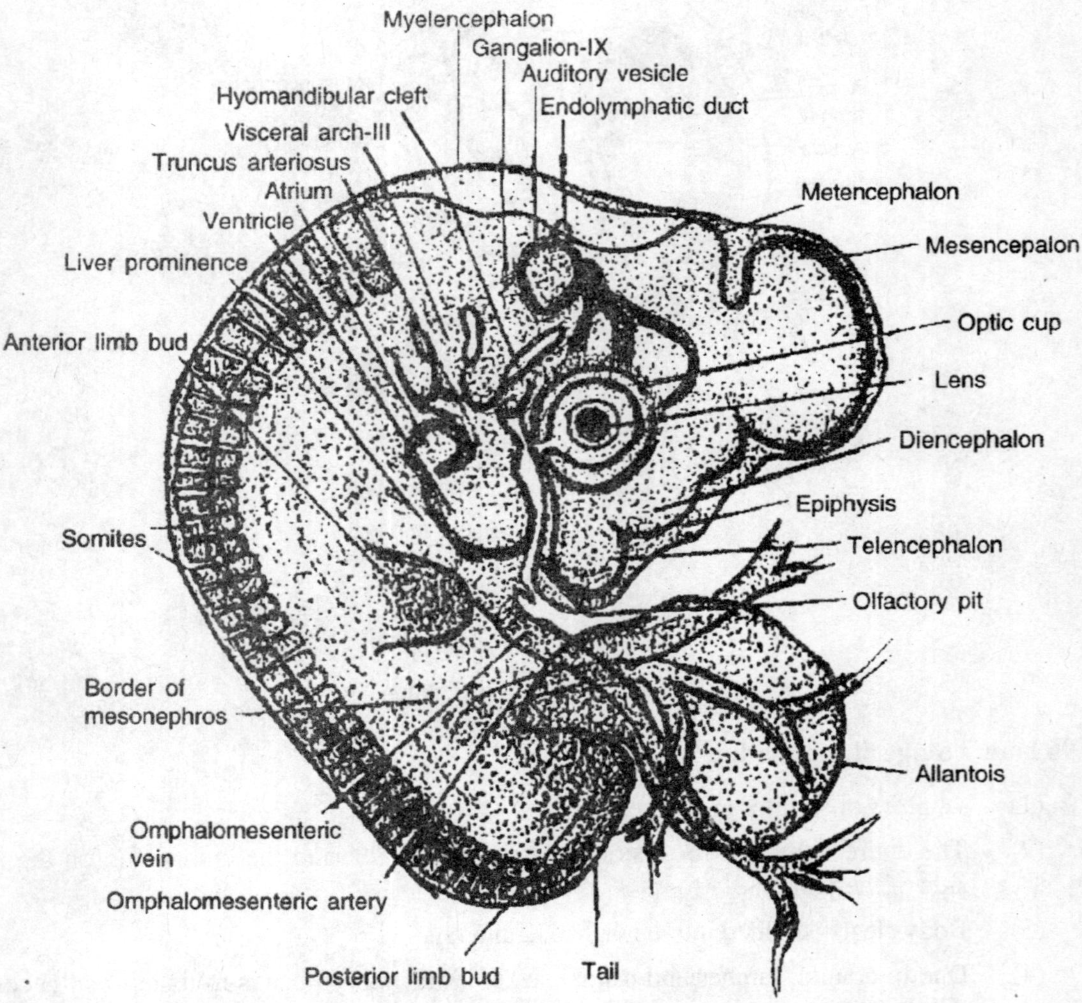

Figure 6.28 (96 hrs of incubation)

1. Demonstration of living embryo after removal of the Blastoderm from the egg

Materials required

Fertilized egg of hen, Enamel tray, Petridish watch glass, Dropper, Washing bottle, Filter paper, Scissors, Bouin's fixative. Incubator and Lithium carbonate.

Demonstration of living embryo after removal of the Blastoderm from the egg

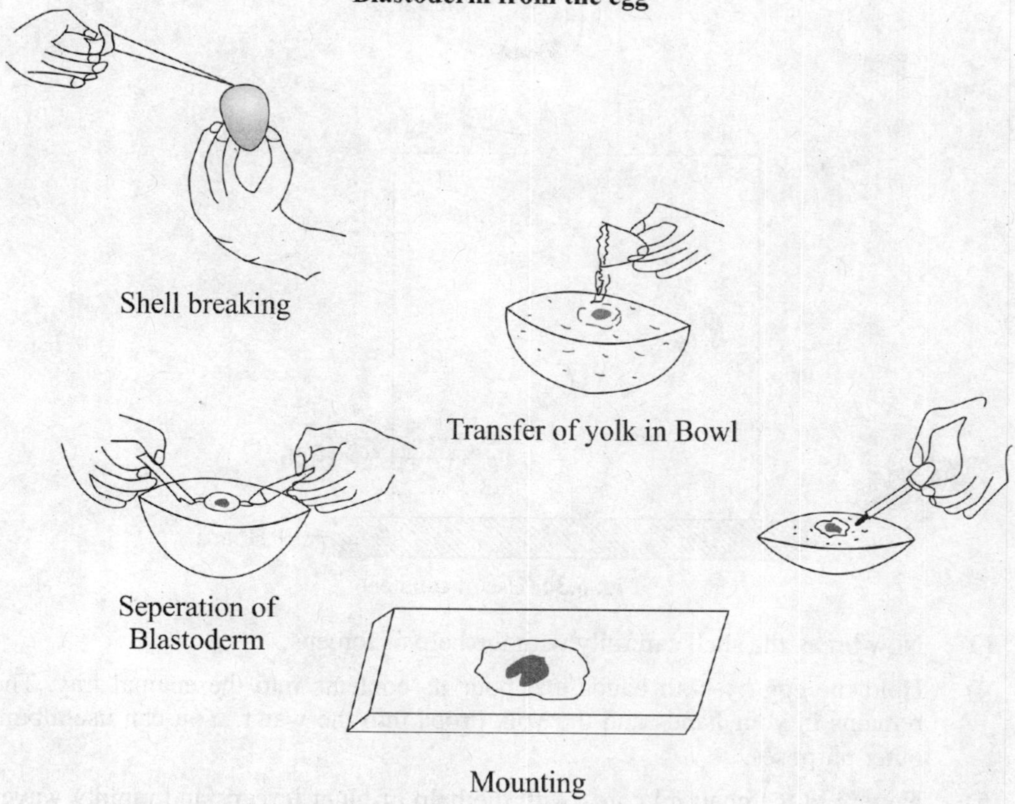

Fig. 6.29 Mounting of blastoderm

Inspection of Fertilized Egg

This is done with the help of a *candler*. It is an apparatus which can be framed by mounting a bulb on wooden board covered by a circular tin. A circular hole is cut on the upper side of the tin so that the egg's pointed end can easily be placed in it leaving the broader end of the egg outside (now the egg is half inside and half outside the hole). Switch on the bulb and view the broader end of the egg from the above.

If the round spot is visible, the egg is not fertilized and discard such eggs for the blastoderm mounting. If any stage of incubation is visible, use it for the experimentation.

Special Remark

16, 18 and 21 hour's stage of incubation are visible as almost round spots therefore look for the area opaca and area pellucida and primitive streak for the confirmation.

Procedure

(1) Fill up the enamal tray (bowl) with water.

(2) Hold the egg in left hand in such a way so that its broader end rest on upper side.

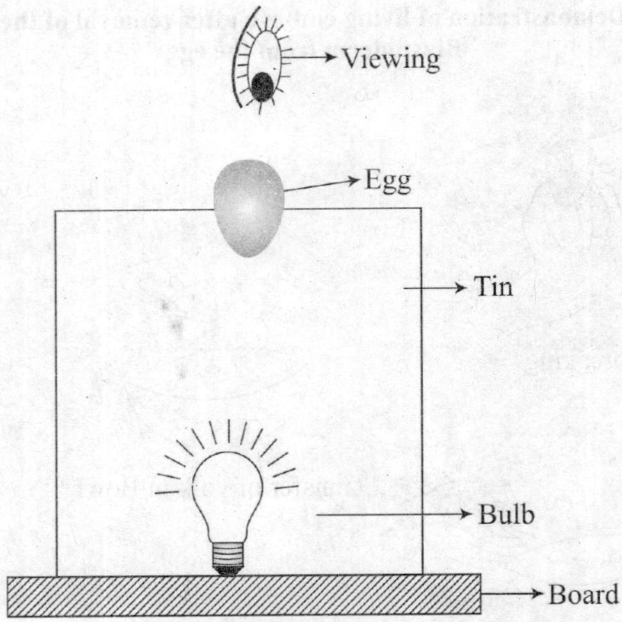

Fig. 6.30 (Use of candler)

(3) Now break the shell carefully with the help of forceps.

(4) Hold the egg by both hands and pour its contents into the enamal tray. The shell remains in your hands and the yolk drops into the water. You can use albumin for other purposes.

(5) Now hold the chalazal cords with the help of blunt forceps and rapidly wave in the water to remove the blastoderm.

(6) Now take out the blastoderm from the enamal tray into a watch glass with the help of a long glass dropper.

(7) Remove the water from the watch glass. Do it patiently otherwise the blastoderm will suck out into the dropper.

(8) Remaining water is suck out with the help of filter paper cuttings. The blastoderm will stick to the watch glass. Spread it carefully and pour few drops of Bouin's fixative.

(9) The blastoderm will be fixed. Excess fixative can be removed by washing with *Lithium carbonate* solution.

(10) Now dehydrate it by passing through the different grades of alcohol (30%, 50%, 70% Eosin, 90%, 100%). Put the blastoderm in xylene for few seconds and mount in DPX. Study the preparation in dissecting microscope and identify it. Record the result in your note book.

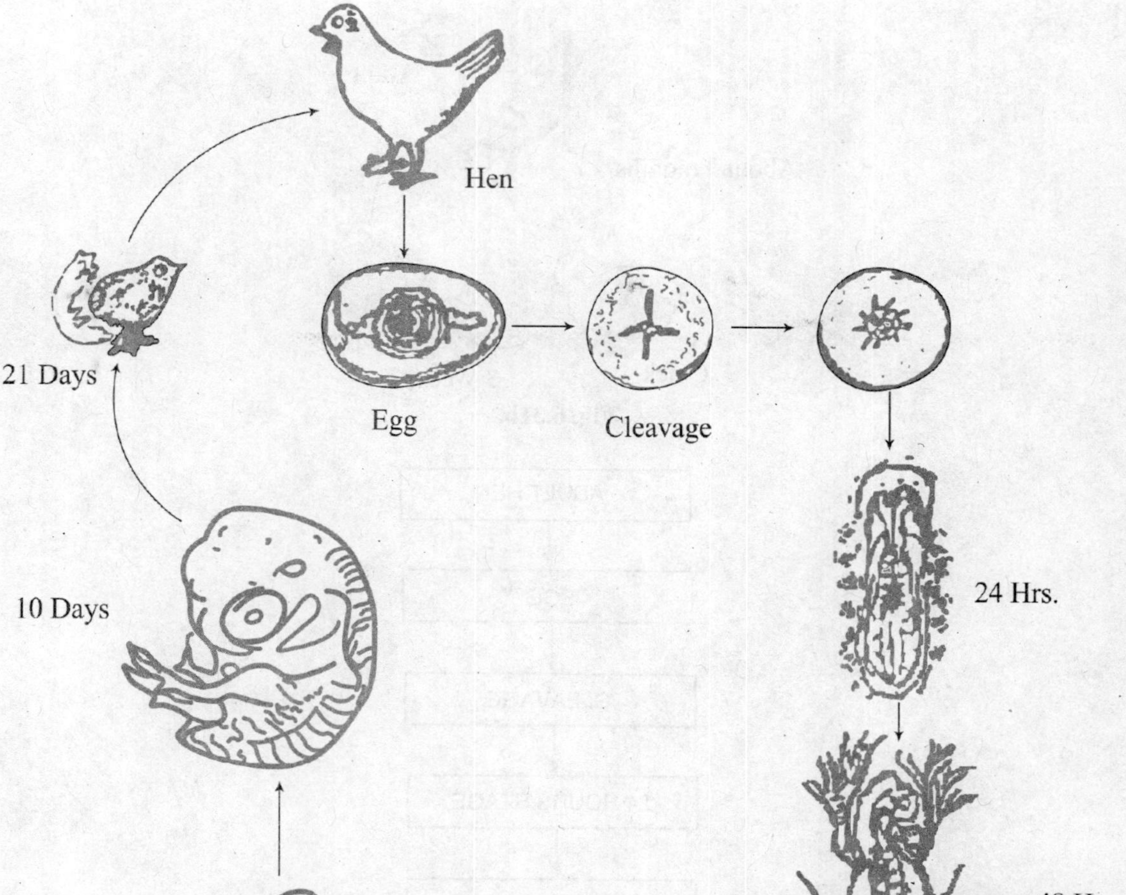

Fig. 6.31a. Life cycle of Hen

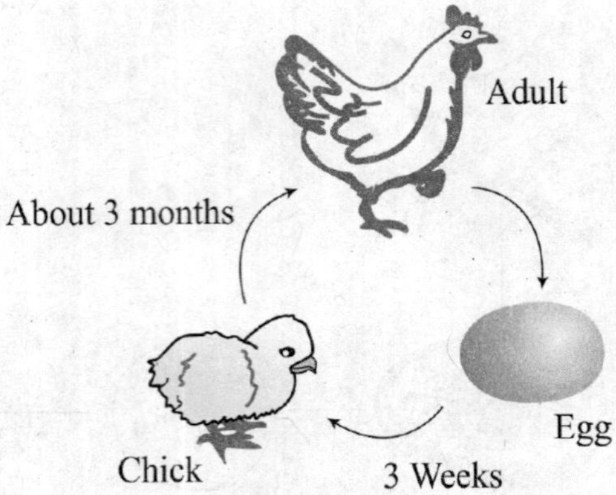

Fig. 6.31b.

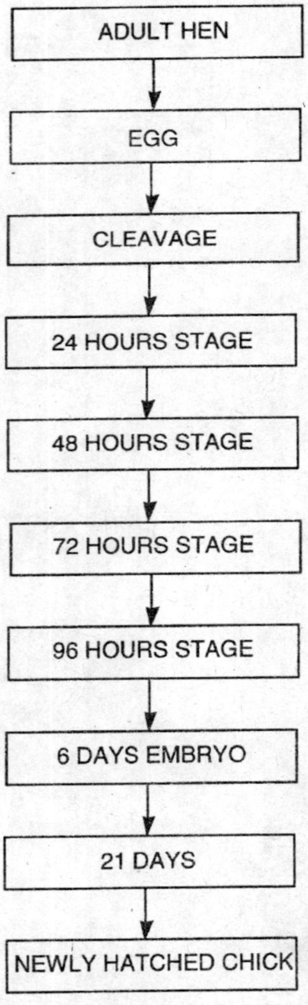

Fig. 6.32. Graphic representation of life cycle of Hen

2. Study of the embryo at various stages of incubation *in vivo* by making a window in the egg shell

Materials required

Fertilized egg, cover slip, saw, transparent tape etc.

Procedure

(1) Procure the fertilized eggs from a poultry farm or any other source and incubate in the incubator at 37ºC in the Laboratory.

(2) Hold the egg in such a way that its broader end remains up (it can also be hold horizontally)

(3) Now make a small square (window) on the shell with the help of a ball point pen and cut the shell with a small saw. Remove the pieces of shell carefully.

(4) View the blastoderm from above and cover the square by a coverslip (either stick with DPX or transparent tape)

(5) Identify the stage and record your results in the note book.

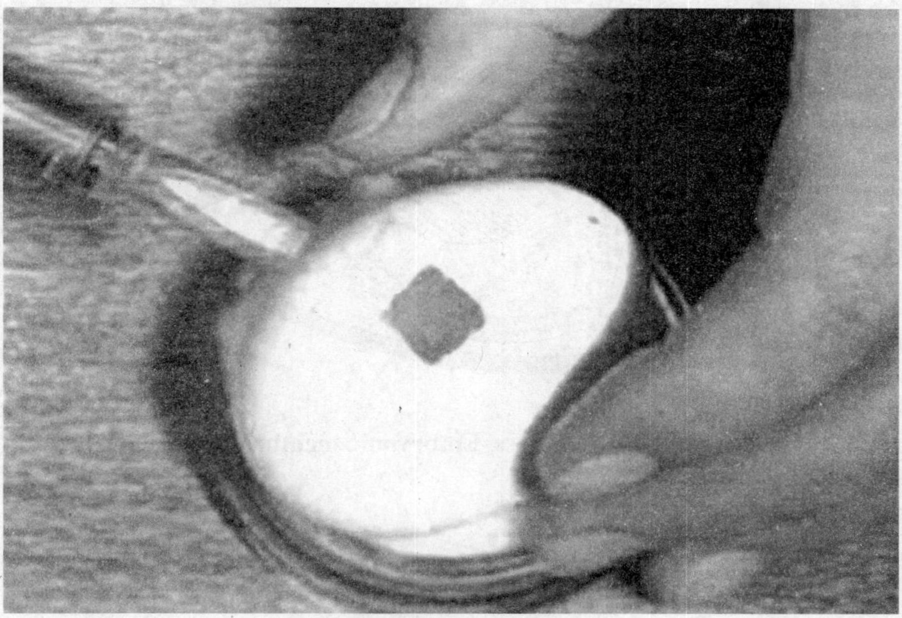

Fig. 6.33. Window preparation

Study of Various foetal envelops in a 10-12 day old chick embryo

Amniotes and Anamniotes are categories of the animals on the basis of foetal membranes. Amniotes possess foetal membranes e.g. Reptiles, Birds and Mammals. Anamniota do not possess foetal membranes eg. Pisces Amphibia. Foetal membranes are also called *extra embryonic membranes*, they do not participate directly in the development of embryo or any structure of embryo rather they play an important role in the development of embryo.

Following four foetal membranes are found in the development of chick.

1. Yolk sac
2. Amnion
3. Chorion
4. Allantois

To study foetal envelops, take a 10-12 day old chick embryo. Remove the shell and study them. Student can take help of the following brief description :

1. Yolk Sac

1. It is the first to develop foetal envelops.
2. *Splanchnopleure* (mesoderm + endoderm) of the area opaca spread over the yolk and covers it, it is called *yolk sac*.
3. It absorbs the yolk and provide nutrition to the developing embryo.

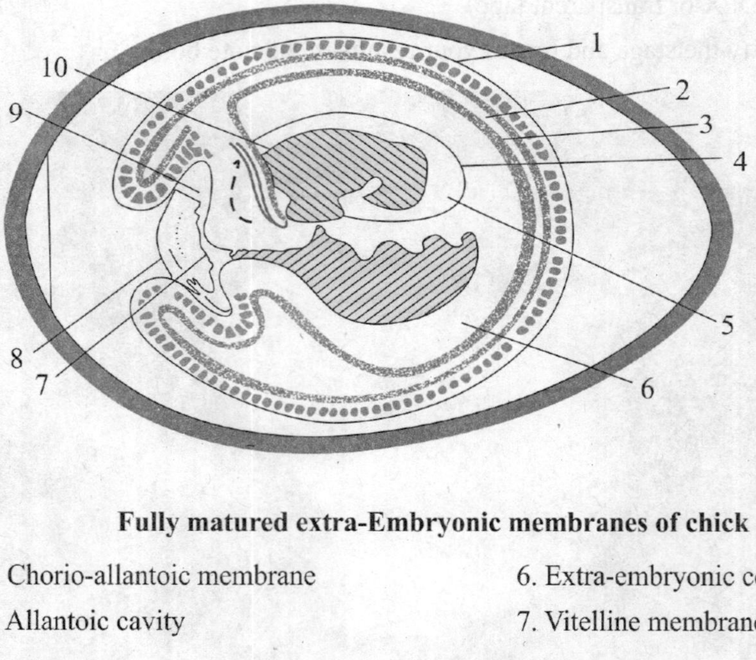

Fully matured extra-Embryonic membranes of chick

1. Chorio-allantoic membrane	6. Extra-embryonic coelom
2. Allantoic cavity	7. Vitelline membrane
3. Allantois	8. Albumen
4. Amnion	9. Albumen sac
5. Amniotic cavity	10. Allantoic stalk

Fig. 6.34: Foetal membranes

2. Amnion and Chorion

1. The first indication of amnion comes at about 30 hours of incubation.
2. It is formed by *somatopleaure* (ectoderm + mesoderm).

3. The outer most layer of somatopleure in which the ectoderm is outside and mesoderm is inside, is called *chorion* (mesoderm is outside and ectoderm is inside in amnion).
4. The cavity between embryo and amnion is called *ammiotic cavity*, a fluid called aamniotic fluid is filled inside it.
5. The cavity between amnion and chorion is called sero amniotic cavity.
6. The amniotic fluid protects the embryo from drying and external shocks.
7. The fluid in sero amniotic cavity provides extra protection to the embryo. It combines with allantois and act as respiratory organ.

Allantois

1. It is developed from the embryonal area.
2. It is made up of *splanchnopleure* (mesoderm + endoderm) where proximal part in the embroyonal area and dirstal grow outside the body of the embryo.
3. It combines with chorion in sero amniotic cavity and forms allanto chorion which act as extra embroyonic lungs.
4. It stores the toxic substances like urea and uric acid (excreted from the protein metabolism) and protects the embryo from toxic effects. It act as excretory organ.

It is thus evident that these foetal envelops play an important role in the development of the embryo.

■■■■

Unit - 7
Biodiversity

1. Appliances used in Biodiversity study

There are many minor equipments and articles needed in conducting surveys to study biodiversity.

Equipments

1. Binocular
2. Camera
3. Spotting telescope
4. Hand/lens
5. GPS
6. Altimeter
7. Pedometer
8. Lux meter
9. Hygrometer/Thermohygrometer

Others

Insect net	Forceps	Paper envelops
Pitfalltrap	Scissors	Plastic envelops
Light trap	Brush, Needle	Wide mouth jar, vials etc.

Binoculars

Many brand names are available eq. Zenith, Sworoveski, Tento, Bushnell, Nikon, Hawk Vanguard, Celestron, Zeiss etc.

The magnification of the binocular is the number is written x. Following features are looked for in a binocular - magnification, objective lens diameter, quality of lens, field of view or exit pupil, quality of lens coating, weight, water proofing and eye strain.

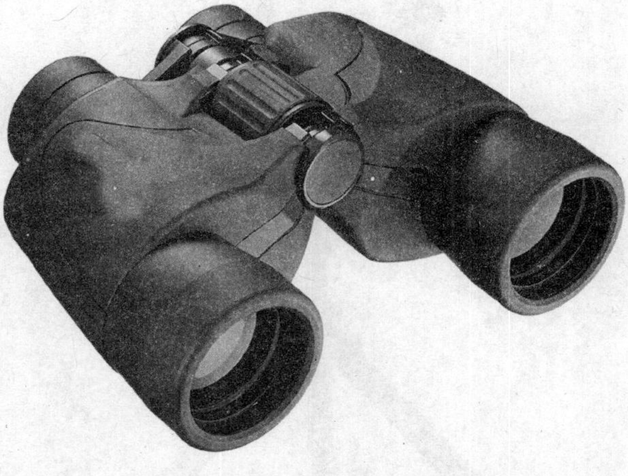

Fig. 7.1a. Bionoculars

A 7 x 50 binoculars means: the second number i.e. 50 is associated with the diameter of objective lens in millimeters so in 7 x 50 binocular "50" means the front lenses of biocular are 50 mm in diameter. Similarly 10 x 50. Here 10 is 10 x magnification and objective lens diameter is 50 mm. The wider the objective lens, the more light the binocular will receive in making the image brighter. 7x to 10x are good.

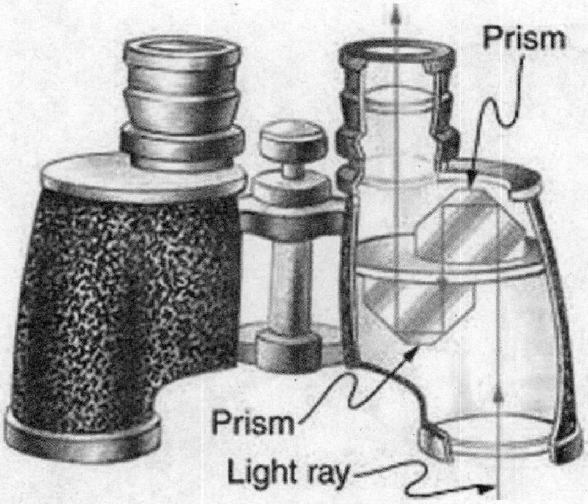

Fig. 7.1b. Working of Binoculars

Camera : Many brands are available. Use the camera available in your Laboratory or now a days android mobile phones are available with very good camera.

Spotting telescope

It is a portable telescope with added optics to present an erect image. Select either 40 x 50 or 20 x 50 or 20 x 70. The number to the left of the `x' always refer the magnification power and number to right of the `x' indicate diameter of objective lens. If it is 7 x 35 means it magnifies objects so they appear 7 times closer.

Fig. 7.2. Spotting telescope.

Hand lens

It is also called magnifying lens. It is easy to carry in fields.

Fig.7.3. Hand lens

GPS

Global Positioning System (GPS) is a satellite based navigation system. It provides geolocation and time information to GPS receiver in all weather conditions, anywhere or near the earth. To note geographical coordinates, altitudes and latitudes, it is used in field observations.

Fig. 7.4. GPS

Altimeter

It is an instrument used to measure the altitude of an object above a fixed level. The measurements are called altimetry which is related to bathymetry. Bathymetry is measurement of depth under water we can also measure pressure as barometer is also there is some altimeters.

Pedometer

One can measure steps taken by a person detecting the motion of the person's hand or hips. Distance travelled by walking or any other means can be measured directly by a GPS receiver. It is used in field observations for the distance measurement in transects etc.

Lux meter

The intensity and amount of the light can be measured. Symbol lx. used in ecological studies and behavioural studies.

Fig. 7.5. Altimeter

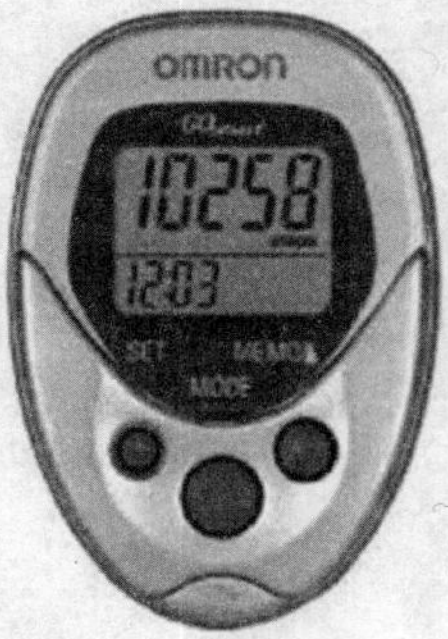

Fig. 7.6. Pedometer

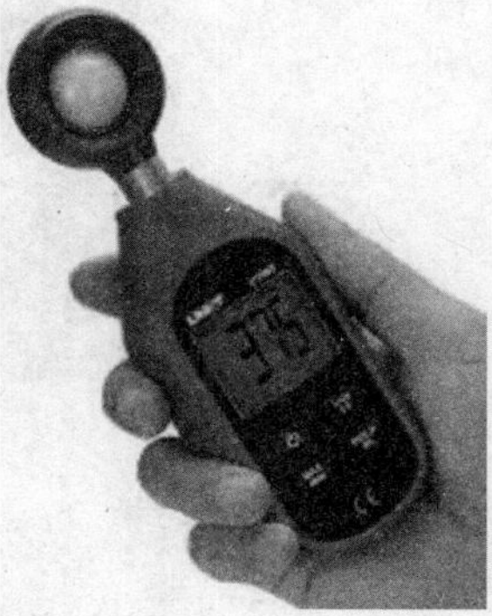

Fig. 7.7. Lux meter

Hygrometer

The water vapours in the atmosphere can be measured by hygrometer. It gives the direct reading or a scale that how much humidity is there in the atmosphere. In various studies the humidity is required. It is useful in other weather based studies.

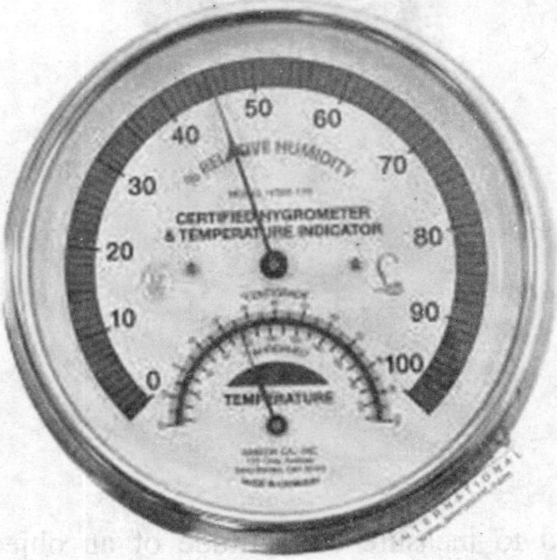

Fig. 7.8. Hygrometer

Insect net

Insect collecting net can have wooden handle or aluminium handle. The diameter is 12" attached to a heavy duty iron ring. Net is 31" deep. It can be of cotton net or plastic net. It is used in sweeping insects during field collections.

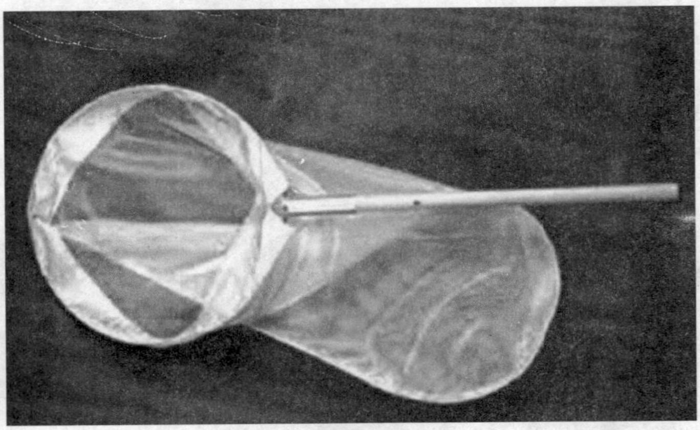

Fig. 7.9. Insect net

Pitfall trap

It is a trapping pit for small animals such as insects. It is used mainly in ecological studies. This device is used to trap insects that are active on ground surface. Pit fall traps usually consist of a beaker or jar that is burried so that the brim of the jar/beaker is level with the ground. The trap is left overnight or stipulated period and inspected later and the data of the catch are recorded.

Fig. 7.10. Pitfall trap

Light trap

It is a device to collect insects during night. A mercury lamp is left lighted for stipulated time, the insects attracted to light are collected in the bottom and counted.

Fig. 7.11. Light trap

2. Nature trails

Make suitable group of students and select a trail either in a Sanctuary or National Park or your college campus. Observe the birds encountered and plants. Make a list of vegetation and Birds.

3. Water sieving

This can be done on a natural pond or lake or stream in your area. Take a sieve (boltine silk) and dip it into water. Collect the animals. Identify them and make a list of each groups. Insects, aquatic beetles, Cyclops, Daphnia, volvox, Euglena, mosquito larvae, aquatic bugs etc. can easily be identified.

4. Biodiversity Survey

Expt. 1

Bird count

Object: To conduct a survey of birds on a site.

Requirements: Bird count data sheet, paper, Pencil and Binocular.

Background information: In the present survey students are required to count the bird species on the site. The sites can be selected by the concerning teacher. It can be a Grassland, Lake, Mountain, Forest, Muddy bank, garden or Monuments or Feeding table.

Methods: The students can be divided into groups for the convenience or will work as a class. It is decided by the teacher as the birds are also to be identified or just to count the different types.

Following methods can be adopted

a. Tansect

In this method the students walk a transect across an area and a suitable distance of 50/100 ft. can be selected. The calls of the individual birds can also be marked as one single bird. The transect method is good for large grassy areas.

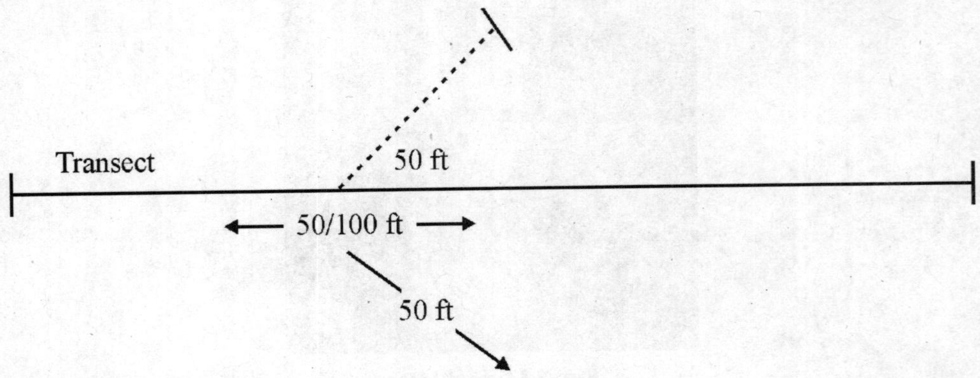

This is belt of 50 ft distance and is called belt counting.

Point sampling

Student walk to a point. Stand there and count all the birds that they see or hear within a radius of certain distance (50 ft). This is good for smaller areas.

Feeding table or feeder station sampling

The feeders can be placed at different areas within the campus or out side. The feeding table can be set or placed out a week ahead of time so that the birds have a chance to find them. The number of food items can be used to attract birds e.g. mix seeds, bajra seeds, sorghum seeds, corn seeds, gram seeds rice or other. This has a draw back as many birds do not come to feeding station.

Data sheet

Date of sampling ____________________

Time of the day ____________________

Weather conditions ____________________

Method of sampling ____________________

Site I	No. of bird types	No. of different calls heard
Site II		
Site III		
Site IV		

Expt. 2

Object: To conduct survey for insect counts on vegetation.

Requirements: Insect count data sheet, paper, pencil, insect net, pitfall trap, vials, Insect box, Forceps, brush.

Background information

In the present survey, students are required to count insects on vegetation in the given area. The study can be conducted at a college garden, or any public park on campus of your college.

Methods

The students can be either divided into groups or taken as class. The sweeping method is applied to sweep insects from the vegetation. During cold months, not much insects can be sweeped. Try pitfall trap method.

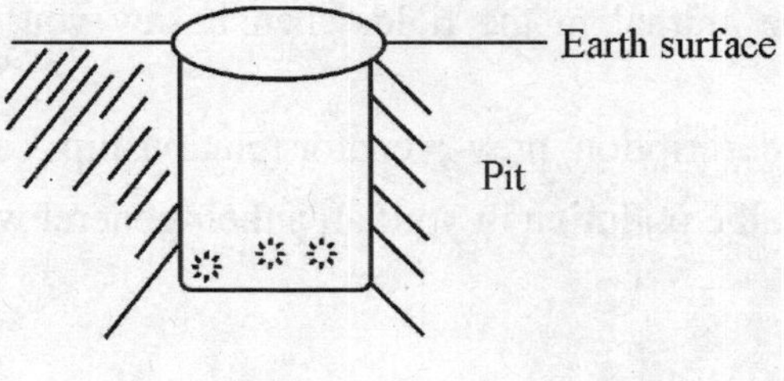

Pitfall trap: A glass or plastic jar (wide mouthed) is fixed in a dig at earth leaving its brim at the level of earth. The crawling insects are fell down in the jar and collected.

The insects are then counted. Different insects are identified and list is prepared.

Insect identification manuals can be used with the be help of your teacher.

Data sheet

Time ____________________

Place ____________________

Weather conditions ____________________

Site I	Sweep net	Pit fall trap	Total no. of insects/ arthropods found
I			
II			

Expt. 3

Object: Discovery hunt in your campus.

Requirements: Notebook, Pen, Pencil, Binoculars, Sweep net, Forceps, Paper envelops etc.

Background information

Biodiversity constitute different type of animals, plants in a given area at a given time. How the ecosystem is stable is depends or its biodiversity. Biodiversity allows wider genetic information therefore, important in ecosystem studies.

Procedure

Divide the students into groups of suitable size and ask them to visit the place allotted to them. It can be any area within the campus or elsewhere in the city including sanctuary, National Park, water body etc. Ask then to observe the area and record their results. e.g. During the visit if they observe a tree then - how tall the tree is, type of canopy, animals or birds taking shelter. Identify the birds and insects observed. Read detail account of the animals observed and find what threads to their survival predominants in the area.

If they come across animals the number of the animals, whether single or groups. What is the reaction/response of the animal in the field when it saw you. Observe the behaviour and record your results.

Other notes on habitat description, prey-predator relationships can also be taken.

Response of animals to the pollution in your area their general well being can be observed.

Survey on vector borne diseases can also be undertaken in the area of your campus and results and observations can be shared with health department.

Activity

Observe three species of animals in your campus.

- Make sketch of all three
- Describe features
- How it utilizes ecosystem.

Interprete your results on a data sheet.

Data Sheet

Species (A)	Species (B)	Species (C)
SKETCH	SKETCH	SKETCH
Description/features	Description/features	Description/features
How it uses ecosystem	How it uses ecosystem	How it uses ecosystem

Expt. 4

5. ANIMAL FACT SHEET

Fact sheets for common wildlife found in your area/state can be prepared. The student can study any five wild animals and keep as record.

The fact sheet can be prepared on the following points.

Systematic position

Phylum ____________________

Class ____________________

Order ____________________

Family ____________________

Zoological Name

Physical characteristics

(i) Height

(ii) Weight

(iii) Head and Body length

(iv) Shoulder height (where applicable)

(v) Tail length

Reproduction and Development

(i) Gestation period

(ii) Litter size

(iii) Weaning

(iv) Sexual maturity

(v) Life span/longevity

Habit and Habitat

(i) Diet

Range/distribution

Predators

Conservation status

(i) IUCN category

(ii) CITES listing

(iii) Threats

(iv) India wildlife Protection Act, 1972

Example

Black buck (Indian Antilop)

Zoological name: *Antilop cervicapra*

Systematic position

Phylum - Chordata

Class - Mammalia

Order - Artiodactyla

Family - Bovidae

Physical characteristics

It stands upto 74 to 84 cm. Male has ringed horns 33-75 cm long

Weight	20-57 kg (male)
	20-33 kg (female)
Speed	50 km/hr

Reproduction and development

Male matures at one and half year and female about 8 months. They generally do not mate before 2 years.

Gestation period - 6 months

Litter size - Single calf is born.

Life span - 10-15 years.

Habit and Habitat

Diet: Herbivorous. Grazes at low grasses, active during the day.

Predators - Pariah dogs Cheetah (not found in India)

Conservation status : Near threatened (NT)

IUCN has mentioned least concerned (LC)

Included in schedule I of wildlife Protection Act, 1972.

In Rajasthan *Taal chhapar* sanctuary in Churu district (Rajasthan) is famous for Black bucks.

Census record : 40,000 India

Fig. 7.12. Black buck

Example - 2

Chital / Spotted deer

Systematic position

Phylum - Chordata

Class - Mammalia

Order - Artiodactyla

Family - Cervidae

Zoological name: *Axis axis*

Physical characteristics

Head and Body length	110-140 cm
Shoulder height	75-100 cm
Tail length	20-30 cm
Weight	45-55 kg (Females)
	65-85 (males)

Antlers in males only. Antlers grow upto 75-80 cm

Reproduction and Development

Sexual maturity - Males at 14 months and females at 9 months

Gestation period - 225-235 days

Litter size - 1

Weaning - youngs fully weaned by 6 months

Life span - 15 years

Fig. 7.13. Spotted deer

Habit and Habitat

Diet - Short grasses, herbs and other foliage.

Predators - Tiger, Leopard, Wild dogs, Wolves, Jackal, Python, Pariah dogs. Active during day (morning and late after noon). Rest in shade during mid day heat.

Conservation status

IUCN, Least Concern (LC), Protected under Wildlife Protection Act, 1972, Schedule 1.

Students can prepare fact sheet on tiger, leopard, lion, rhesus monkey, langur, bear, otter wolf, jackal, fox, elephant, etc.

■■■■

Appendix - I

1. Schiff's reagent

Basic fuchsin - 2 gm

Potassium metabhisulphite 3.8 gm

($K_2S_2O_5$)

0.15 NHCl - 200 ml

Dissolve the stain fuchsin into metabisulphite and HCl. Shake well at the interval of 2 hrs. It is straw coloured. Add 500 mg activated charcoal and filter. Store in dark bottle in Refrigerator.

2. Sulphite wash

10% sodium metabisulphite	-	30 ml
1N Hcl	-	30 ml
Distilled water	-	500 ml

Add metabisulphite into distilled water and then add acid.

3. Carnoy's fixative

Absolute alcohol (ethyl)	-	60 ml
Glacial acetic acid	-	10 ml
Chloroform	-	30 ml

(Fix the tissue at least 15-60 minutes)

4. Fast Green

0.01% in 95% alcohol or 1 gm fast green in 95% alcohol

5. Bouin's fixative

Picric acid	-	75 ml
40% HCHO	-	25 ml
Glacial acetic acid	-	5 ml

6. FAA

40% HCHO	-	10 ml
70% ethyl alcohol	-	85 ml
Glacial acetic acid	-	5 ml

7. Normal saline

NaCl	-	0.78 gm
Distilled water	-	100 ml

8. Leishman stain

Leishman stain	-	0.15 gm
absolute alcohol	-	100 ml

Allow to ripe for 48 hours. It is good for protozoans & blood cells.

9. Congo red

congo red	-	0.5 gm
distilled water	-	100 ml

10. Crystal violet

Crystal violet	-	2 gm
ethyl alcohol (90%)	-	20 ml
distilled water	-	180 ml

11. Giesma stain

Giesma powder	-	0.5 gm
glycerine	-	33 ml
absolute alcohol (methyl)	-	33 ml

Keep the mixture over night for incubation at 60ºC and filter.

Note : Giesma is a mixture of methylene blue, Azure A and B staining solution is made by diluting stock solution.

(2.5 ml stock solution + 1 ml methanol + 50 ml Giesma water (pH 6.9).

12. Methylene Blue

Methylene blue	-	1.5 gm
ethyl alcohol (90%)	-	100 ml

13. Zenker's solution

$K_2Cr_2O_2$ - 2.5 gm

$HgCl_2$ (Mercuric chloride) - 15 gm

Distilled Water (DW) - 100 ml

Sodium sulphate - 1 gm

Glacial acetic acid - 5 ml

14. Karl's Fluid -

Ethyl alcohol 90% - 170 ml

40% HCHO - 60 ml

DW - 280 ml

Glacial acetic acid - 20 ml

(Add acetic acid just before using)

15. Neutral red

1 part neutral red + 3000 parts ethyl alcohol

16. Iodine (Lugol's)

Kl - 10 gm

DW - 100 ml Dissolve and add iodine

Iodine - 5 gm

(It is good for cilia, flagella & nucleus

Aquous Iodine

Kl - 2 gm

DW - 100 ml

Iodine Crystal - 1 gm

17. Picro Indigo carmine

Picric acid (saturated solution in 90% alcohol) - 1 part

Indigo solution in 70% alcohol - 2 parts

(It is used for chitinous material)

18. Kleinberg's fixative

Picric acid + 2% aquous solution of H_2SO_4.

19. Allen's Fluid

Saturated picric acid aquous 75ml, Formalin 15 ml, Acetic acid 10 ml and urea 1 gm.

20. Acid Alcohol

Conc. HCl 100 ml and Isopropyl alcohol 200 ml.

■■■■

Appendix - II

PRACTICAL SYLLABUS OF RAJASTHAN UNIVERSITY, JAIPUR

Practical - Zoology

Min. Marks : 18 | 4 Hrs./Week | Max. Marks : 50

I. Microscopic Techniques:

1. Organization and working of Optical Microscope: Dissecting and compound microscopes.
2. General methods of microscopic slide preparation; narcotization; fixing and preservation; washing; staining; destaining; dehydration; clearing and mounting.
3. General idea of composition, preparation and use of:
   - (i) Fixatives: Formalin, Bouin's fluid.
   - (ii) Stains: Aceto-carmine, Aceto-orcein, Haematoxylin, Eosin
   - (iii) Common reagents: Normal saline, Acid water, Acid alcohol and Mayer's albumin.
4. Collection and Culture Methods:
   - (i) Collection of animals from their natural habitat during field trips such as *Amoeba, Paramecium, Euglena, Daphnia, Cyclops* etc.
   - (ii) Culture of *Paramecium* in the laboratory and study of its structure, life - processes and behaviour in life state.

II. Study of Microscopic Slides and Museum Specimens:

Protozoa: *Amoeba, Euglena, Trypanosoma, Giardia, Entamoeba, Elphidium (Polystomella), Foraminiferous shells, Monocystis,Plasmodium, Parameciu, Leishmania, Paramecium* showing binary fission and conjugation, *Opalina, Nyctotherus, Balantidium, Vorticella.*

Porifera: *Leucosolenia, Euplectella, Spongilla,* T.S. Sycon, Spicules, Spongin fibers, Gemmules.

Coelenterata: *Millepora, Physalia, Velella, Aurelia, Alcyonium, Gorgonia, Pennatula,* Sea anemone, Stone corals, *Obelia* colony and medusa.

Ctenophora: Any Ctenophore

Platyhelminthes: *Taenia, Planaria,* Fasciola (WM), T.S. body of *Fasciola,* Miracidium, Sporocyst, Redia and Cercaria Larvae of *Fasciola,* Scolex, T.S. mature proglottid of Taenia, gravid proglottid, Cysticercus larva.

Aschelminthes: *Ascaris, Wuchereria, Dracunculus.*

Annelida: Neries, Heteroneries.

III. Anatomy

Earthworms: External features, general viscera, alimentary canal, reproductive system and nervous system.

Leech: External features, alimentary canal, reproductive and nervous system.

IV. Study of the following through Permanent Slide Preparation: *Paramecium, Euglena,* Foraminiferous shells, Sponge spicules, Spongin fibres, Gemmule, *Hydra, Obelia* colony and Medusa; Parapodium of Nereis and heteronereis, Leech : testicular nephridia.

V. Exercises in Cell Biology

1. Squash preparation for the study of mitosis in onion root tip, permanent slides of mitosis (all stages).
2. Squash preparation for the study of meiosis in grasshopper or cockroach testes, permanent slide of meiosis (all stages).
3. Study of giant chromosomes in salivary glands of *Chironomous* or *Drosophila* larva.
4. Study of cell permeability using mammalian R.B.C.

VI. Exercises in Genetics:

A. **Study of *Drosophila:***
   1. Life cycle and an idea about its culture.
   2. Identification of male and female.
   3. Identification of wild and mutants (yellow body, ebony, vestigial wing and white eye).
   4. Study of permanent prepared slides: Sex comb and salivary gland chromosomes.

B. **Numerical problems based on monohybrid and dihybrid cross.**

C. **Identification of blood groups (A, B, O & Rh factor).**

VII. Developmental Biology

1. **Study of development of frog/toad with the help of Charts/Slides/Models:**
   (i) Eggs, cleavage, blastula, gastrula, neurula, tail-bud, hatching, mature tadpole larvae, metamorphic stages, toadlet/froglet.

(ii) Histological slides: Cleavage, blastula, gastrula, neurula and tail-bud stage.

2. **Study of development of chick with the help of whole mounts/Charts/Slides/ Modes**

(i) 18 hrs, 21 hrs, 24 hrs, 33 hrs, 48 hrs, 72 hrs and 96 hrs of incubation.

(ii) Primitive streak stage in living embryo, if possible, after removal of the blastoderm from the egg.

(iii) Study of the embryo at various stages of incubation *in vitro* by making a window in the egg-shell may also be demonstrated.

(iv) Study of various foetal membranes in a 10-12 day old chick embryo.

Scheme of Practical Examination and Distribution of Marks

Time : 4 Hrs | Min Pass Marks: 18 | Max. Marks: 50

		Regular	Ex./N.C. Students
1	Anatomy (any system)	6	6
2	Permanent Preparation	4	6
3	Cell Biology and Genetics	4 + 4	6 + 6
4	Developmental Biology	6	5
5	Identification and comments on Spots (1 to 8)	16	16
6	Viva Voce	5	5
7	Class Record	5	–
		50	**50**

Notes:

1. With reference to anatomy (Dissection), study of prescribed types and developmental biology (charts/models) candidates must be well versed in the study of various system. CD ROMs multimedia computer based simulations including computer assisted learning (CAL) and other soft wares may be used.

2. With reference to permanent preparations and microscopic slides, in case of non-availability, the exercise should be **substituted with diagrams, photographs, models, charts, etc.**

3. Candidates must keep a record of all work done in the practical class and submit the same for inspection at the time of the practical examination.

4. The candidates may be asked to write detailed methodology wherever necessary and separate marks may be allocated for the same.

5. Mounting material for permanent preparations would be as per the syllabus or as available through collection and culture methods.

6. **It should be ensured that animals used in the practical exercises are not covered under the wild life act 1972 and amendments made subsequently.**

■■■■

Appendix - III
SYLLABUS of M.S. Braj University, Bharatpur

First Year Practical Zoology

Min. Marks : 18 **4 Hrs. duration** **Max. Marks : 50**

I. Microscopic Technique

1. Organization and working of optical microscopes : dissecting and compound microscope.
2. General methods of microscopical permanent preparations : Narcotization; fixing and preservation; washing; staining, destaining, dehydration; clearing and dealcoholization; mounting. General idea of composition, preparation and use of :
   (i) Fixatives : Formalin, Bouin's fluid.
   (ii) Stains : Borax carmine, acetocarmine, acetoorcein, haematoxyline, eosin.
   (iii) Common reagents : Normal saline, Ringer's solution, acid water, acid alcohol and Mayer's albumin.
3. Collection and Culture Methods.
   (i) Collection of animals from their natural habitat during field trips, e.g., Amoeba, Paramecium, Euglena, Planaria, Earthworm, Daphnia, Cyclops, etc.
   (ii) Culture of Paramecium in the laboratory and study of its structure, life processes and behaviour in the living state.
   (iii) Collection and study of invertebrate parasites : Rectal Ciliates of frog.
   (iv) Vermicomposting : Theory & Practice.

II. Study of Microscopic Slides and Museum Specimens

Slides of Histology : All the tissues specified in theory.

Protozoa : Euglena, Trypanosoma, Giardia, Entamoeba, Elphidium (Polystomella), Amoeba, Foraminiferous shell, Monocystis, Plasmodium, Paramecium, Paramecium showing binary fission and conjugation, Opalina, Nyctotherus, Balantidium, Vorticella.

Porifera : Leucosolenia, Euplectella, Spongilla, L.S. and T.S. Sycon, spicules, spongin fibres, gemmules.

Coelenterate : Millepora, Physalia, Vellela, Aurelia, Alcyonium, Gorgonia, Pennatula, Sea anemone, stone corals, Obelia colony and medusa.

Ctenophora : Any ctenophore.

Platyhelminthese: Taenia, Planaria, Fasciola (W.M.), T.S. body of Fasciola through various regions, Miracidium, Sporocyst, Radia, Cerlaria Metacercaria, Scolex, T.S. mature proglothid of Taenia, Cysticoreus larva.

Aschelminthese : Ascaris, Wuchereria, Dracurnculus

III. Permanent Preparation and Study of the following :

Paramecium, Euglena, rectal ciliates of frog, forminiferous shells, sponge spicules, spongin fibres, gemmule, Hydra, Obelia colony and medusa, cyclops, Daphnia.

Parapodium of Nereis and Heteronereis, ovary, nephridia, nerve ring and setae of earthworm; salivary glands and trachea of cockroach.

IV. Exercises in Cell Biology

1. Squash preparation for the study of mitosis in onion root tip.
2. Sqash preparation for the study of meiosis in grasshopper or cockroach testes.
3. Study of giant chromosomes in salivary glands of Chironomus or Drosophila larva.
4. Study of cell permeability using mammalian RBCs.

V. Exercises in Genetics

1. Study of Drosophila: Life cycle and an idea about its culture.
2. Identification of male and female Drosophila.
3. Identification of wild and mutant (yellow body, ebony body, vestigial wings, white eye and vestigial ebony mutants) Drosophila.
4. Study of permanent prepared slides : Drosophila : sex comb, salivary glands chromosomes.
5. Identification of blood groups (A, B, O & Rh Factor).

VI. Developmental Biology

1. Study of development of frog/toad with the help of :
   (i) Preserved materials available : Eggs, cleavage, blastula, gastrula, neurula, tail bud, hatching, mature tadpole larvae, metamorphic stages, toadlet/froglet.
   (ii) Histological slides : Cleavage, blastula, gastrula, neurula and tail-bud stage.
2. Study of development of chick with the help of
   (i) Whole mounts : 18 hrs, 21 hrs, 24 hrs, 33 hrs, 72 hrs and 96 hrs of incubation period embryos.

(ii) If possible, primitive streak stage in living embryo after removal of the blastoderm from the egg may be demonstrated.

(iii) Study of the embryo at various stages of incubation *in vivo* by making a window in the egg shell may also be demonstrated.

(iv) Study of various foetal envelopes in a 10-12 day old chick embryo (amnion, chorion, allantois, yolk sac).

VII. Biodiversity

(i) Appliances used in Biodiversity study.

(ii) Nature trails, water sieving.

(iii) Discovery hunt in college campus/university campus/Forest reserve/Sanctuaries/National Parks.

(iv) Biodiversity survey: Insect count on vegetation, Birds count with general information on survey methods.

(v) Preparation of fact sheet of common wildlife found in your area/campus. Field visit/Excursion to wildlife areas.

Scheme of Practical Examination and Distribution of Marks of Maharaja Surajmal Braj University, Bharatpur

Time : 4 hrs **Min. Pass Marks : 18** **Max Marks : 50**

	Regular	**Ex-students**
1. Study of Biodiversity	06	4
2. Permanent Preparation	04	7
3. Exercise in Cell Biology and Genetics	4 +4	6 + 6
4. Exercise in Developmental Biology	6	6
5. Identification and Comments on Spots (1 to 8)	16	16
6. Viva Voce	5	5
7. Class Record	5	
Total	50	50

Notes :

1. With reference to dissection, candidates must be well versed with the technique of flag labelling and black paper insertion as the case may be for a clear illustration.
2. With reference to whole mounts and museum specimens, in case of unavailability of certain animals types, diagrams, photographs, models, etc. should be substituted. Study will include classification (upto classes) with diagnostic characters and comments.
3. Candidates will keep a record of all work done in the practical class and it will be submitted for inspection at the time of the practical examination.
4. Mounting material will be provided as per the syllabus or available from Collection and Culture Methods.
5. They detailed methodology may be asked to write wherever necessary and separate marks may be allocated for it.
6. It should be ensured that animal used in practical exercises are not covered under the Wildlife Act 1972 and amendments made subsequently.

■■■■

References

Agarwal, S.K. and Niazi, I.A. *Normal table of developmental stages of Indian bull frog. Rana tigerina* (Daud.) Nat. Acad. Science, Allahabad.

Balinsky, B.I. *An Introduction to Embryology.* W.B. Saunders, Philadelphia & London, 1971. Bernil N.J. Developmental Biology, McGraw Hill.

Bisnal, M.S. and Kanamadi, R.D. *Normal developmental stages of the Toad, Bufo melanostictus: A systematic study in relation to time.*

Freeman, W.H. and Bracegirdle, B. *An Atlas of Embryology.* ELBS, London, 1976.

Goel, S.C. *Principles of Animal Developmental Biology.* HPH Bombay, 1984.

Jain, P.C. *Chordate Embryology.* Vishal Publications, Jalandhar City, India.

Majumdar, N.N. *Text Book of Vertebrate Embryology.* Tata Mc-Graw Hill, New Delhi, 1990.

Nair, P.K.G., Achar, K.P. and Trigunayat, M.M. *Elements of Chordate Embryology.* Pragati Prakashan, Meerut, 1997.

Mathews, W.W. *Atlas of Descriptive Embryology.* McMillan, London, 1976.

McEven, R.S. *Vertebrate Embryology.* McGraw Hill Book Co. Ltd., New York, 1949.

Mohanty, A.K. and Dutta, S.K. *Breeding and Development of Linonectes limnocharis* (Anura-Ranidae).

Rao, K.V. *Developmental Biology: A Modern Synthesis.* Oxford & IBH, Delhi, 1994.

Rugh, R.A. *Manual of Experimental Embryology,* New Delhi, 1962.

Swarup, H., Arora, S. and Pathak, S.C. *Laboratory Techniques in Modern Biology,* Kalyani Publishers, New Delhi, 1990.

Trigunayat, M.M. and Singh, N.P. *Laboratory Studies on the Development of Lasioderma serricorne.* Natl. Symp. on Developmental Biology, 21-23 Feb. 1997, Orissa.

Blood Groups in Man : R.R. Race and R. Sanger.